Molecular Phylogeny, Biogeography and an e-Monograph of the Papaya Family (Caricaceae) as an Example of Taxonomy in the Electronic Age

Fernanda Antunes Carvalho

Molecular Phylogeny, Biogeography and an e-Monograph of the Papaya Family (Caricaceae) as an Example of Taxonomy in the Electronic Age

With a foreword by Prof. Dr. Susanne S. Renner

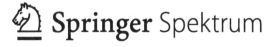 Springer Spektrum

Fernanda Antunes Carvalho
Munich, Germany

Dissertation Ludwig-Maximilians-Universität München, Germany, 2014

ISBN 978-3-658-10266-1   ISBN 978-3-658-10267-8 (eBook)
DOI 10.1007/978-3-658-10267-8

Library of Congress Control Number: 2015941425

Springer Spektrum

Printed on acid-free paper

Springer Spektrum is a brand of Springer Fachmedien Wiesbaden
Springer Fachmedien Wiesbaden is part of Springer Science+Business Media
(www.springer.com)

Para Bia, Quincas e Tânia

Foreword by the Supervisor

In the flowering plants, there are currently 3–4 synonyms for every accepted name (http://www.theplantlist.org). This fact is not a harmless nuisance. Synonymous names cause at least two kinds of errors: they result in wrong assumptions about species' geographic ranges, and they make it difficult to find information about species because users cannot know which name refers to what. The judgment about what is a good biological species and which names are synonyms is made during monographic research. Such research consists in evaluating the information pertaining to all names published for the species, subspecies, or other forms in a genus or family, along with studying all specimens to which these names have been applied (rightly or wrongly). The very best monographs also include geographic distribution maps and DNA sequences from specimens representing the group's species. Based on these data, a monographer reaches a conclusion about which names refer to true biological species. He or she then constructs a key to identify the accepted species and prepares authoritative species descriptions and maps.

Fernanda Antunes Carvalho carried out monographic research on the papaya family, Caricaceae, between April 2010 and December 2013. Her dissertation consisted of several chapters, three of which are included in the present book. A novel aspect of Dr. Carvalho's monograph of the Caricaceae was that she availed herself of the cybertaxonomy platform 'Botanical Research and Herbarium Management System' (BRAHMS) to also develop an online version of her monograph that will be updated at certain intervals. In her work, she allocated the about 200 names relevant to the family Caricaceae to 34 biological species and one hybrid. To arrive at this conclusion Dr. Carvalho studied of about 2000 herbarium collections as well as relevant type specimens (specimens on which particular names are based).

Fernanda Antunes Carvalho also used molecular-phylogenetic methods to infer the phylogenetic relationships in the papaya family. Her phylogenetic work in 2011/2012 led to the discovery that the closest relatives of papaya (*Carica papaya*) all live in Mexico and Guatemala, not in the Andes as often suggested. This discovery fits with other data that likewise indicate that the papaya was domesticated in Mexico/Guatemala.

The book that Fernanda Antunes Carvalho is presenting here brings together all information that is available today about the evolution, distribution, ecology, morphology, and conservation status of the species in the papaya family. Together with the online data and image database that she maintains about this family, it forms the foundation of all current and future research on this family.

Munich, March 2015   Prof. Dr. Susanne S. Renner

Declaration of Contribution

In this thesis, I present the results of my doctoral research, carried out in Munich from April 2010 to December 2013 under the guidance of Prof. Susanne S. Renner. My thesis resulted in four manuscripts of which two are published (Chaps. III and IV), one is in review (Chap. II) and the fourth (Chap. V) has yet to be submitted. In addition, I worked on an electronic monograph of Caricaceae available at http://herbaria.plants.ox.ac.uk/bol/caricaceae, two book chapters and a tutorial, and gave the talks listed below. I generated all data and conducted all analyses myself, except for the karyological analyses (Chap. V), which were done in collaboration with Alexander Rockinger, Martina Silber, and Aretuza Sousa. Writing and discussion of all manuscripts involved collaboration with Prof. Dr. Susanne Renner.

List of publications

Carvalho FA, Renner SS (2012) A dated phylogeny of the papaya family (Caricaceae) reveals the crop's closest relatives and the family's biogeographic history. *Molecular Phylogenetics and Evolution* 65: 46–53

Carvalho FA (2013 onwards) e-Monograph of Caricaceae. Vers. 1, Nov. 2013. [Database continuously updated]. http://herbaria.plants.ox.ac.uk/bol/caricaceae

Carvalho FA, Filer D, Hopkins M (2013) Using images to enter data in BRAHMS. Tutorial available for download at: http://herbaria.plants.ox.ac.uk/bol/brahms/GroupResources

Carvalho FA, Renner SS (2013) The phylogeny of Caricaceae. *In:* Ming R, Moore PH (eds) *Genetics and Genomics of Papaya.* Springer, Heidelberg, New York, pp. 81–82

Carvalho FA, Renner SS (2013) Correct names for some of the closest relatives of *Carica papaya*: A review of the Mexican/Guatemalan genera *Jarilla* and *Horovitzia*. *PhytoKeys* 29: 63–74

Carvalho FA, Filer D, Renner SS (2014) Taxonomy in the electronic age and an e-monograph of the papaya family (Caricaceae) as an example. *Cladistics* (published online 13 Aug. 2014), DOI: 10.1111/cla.12095

Carvalho FA (2015) Caricaceae. *In:* Davidse G, Sousa M, Knapp S, Chiang Cabrera F & Ulloa Ulloa C (eds) Flora Mesoamericana Vol. 2, Parte 3: Saururaceae a Zygophyllaceae, Missouri Botanical Garden Press, Monsanto.

Oral Presentations

Carvalho FA (2013) Biogeography, systematic, and cyber-monograph of the papaya family (Caricaceae) *EES^LMU Conference 2013*, Ludwig Maximilians University, Munich, Germany, October 8–9, 2013 (c. 20 min)

Carvalho FA (2013) BRAHMS (Botanical Research and Herbarium Management System). Introduction and hands-on. *Workshop of the Systematics and Biodiversity group and Alexandre Antonelli's research group at the Department of Biological and Environmental Sciences, University of Gothenburg*, Gothenburg, Sweden, August 19–30, 2013 (one full day workshop)

Carvalho FA, Renner SS (2013) Climatic niche divergence in old sister lineage splits of Caricaceae, but not young species pairs. *BioSyst.EU 2013 Global systematics!* Vienna, Austria February 18–22, 2013 (c. 20 min)

Carvalho FA (2012) Evolution and biogeography of Caricaceae and the closest relatives of papaya. Invited talk in the *Instituto de Biología, Universidade Autónoma de México*, Mexico City, Mexico Aug. 7, 2012 (c. 40 min)

Carvalho FA, Renner SS (2011) The papaya tree (*Carica papaya*) belongs in an herbaceous Meso-american clade. *BioSystematics Berlin 2011*. Berlin, Germany, February 21–27, 2011 (c. 15 min)

Carvalho FA (2010) Una introduction sobre BRAHMS (Botanical Research and Herbarium Management System), *Instituto de Botánica Agrícola, Universidade Central de Venezuela*, Maracay, Venezuela March 26, 2010 (c. 60 min)

Herbaria Visited

- BHCB, MBM, UPCB, R, RB, HB, VEN, MY, MEXU, GUADA, IBUG, K, BM, OXF, GB, W, WU, B, M. More than 1000 additional specimens borrowed (studied and annotated) from HUEFS, F, and NY

Field Work

- Paraná, Brazil, January 2012; Mérida, Venezuela, March 2012; Jalisco and Oaxaca, Mexico, August 2012

Funding

- *Conselho Nacional de Pesquisas* (CNPq 290009/2009-0) provided a 4-year scholarship (April 1, 2010 – March 31, 2014)
- *Deutsche Forschungsgemeinschaft* (DFG RE 603/13) funded the project entitled "A cybermonograph and phylogeny of the papaya family, Caricaceae: providing the context for the fully sequenced genome of a worldwide crop"
- EES^LMU Travel Grant provided financial support to present talks at conferences in Berlin and Vienna. I also visited herbaria in these cities
- Ray Ming supported my fieldwork in Mexico in 2012 with part of his grant from the *U.S. National Science Foundation* (NSF) Plant Genome Research Program (Award No. DBI-0922545)

Summary

This dissertation addresses an issue of key importance to the field of systematics, namely how to foster taxonomic work and the dissemination of knowledge about species by taking full advantage of electronic data and bioinformatic tools. I tested and applied modern systematic tools to produce an electronic monograph of a family of flowering plants, Caricaceae. In addition to a taxonomic revision, a molecular phylogeny of the family that includes representatives of all biological species clarifies the evolutionary relationships. Based on the plastid and nuclear DNA data, I inferred historical processes that may have shaped the evolution of the Caricaceae and explain their current geographic distribution.

The first part of my thesis focuses on the development of an electronic monograph using existing infrastructures of Information Technology (IT) and bioinformatic tools that together set the stage for a new era of systematics. I address the problem of synonyms and the importance of taxonomic monographs as the portal for the entire information available about species, including all names published since 1753. Using relatively cheap gadgets (a small digital camera and a portable digital microscope), I rather efficiently gathered data from herbarium specimens and organized these data in a dynamically updated electronic monograph of Caricaceae, using the Botanical Research and Herbarium Management System (BRAHMS) developed at the University of Oxford. The e-monograph includes distribution maps (based on 2201 georeferenced collections), photos of 3943 herbarium specimens (and weblinks to high resolution images of type specimens), highly detailed plates illustrating all species, as well as comprehensive data on morphology, chromosome numbers, phenology, uses, and habitat. I revised all extant 233 names, solving nomenclatural and typification problems, and built multi-access identification keys for all species and genera using Xper2, developed at the Université Pierre et Marie Curie in Paris.

The second part of my thesis focuses on the phylogeny and biogeography of Caricaceae. I produced the first complete DNA-based phylogeny of the family including all genera and accepted species and discovered that the closest relatives of papaya are four species endemic to Mexico, Guatemala, and El Salvador. Together with the current distribution of the wild form of papaya (which has smaller and harder fruits than the cultivated form), the phylogeny supports the idea that papaya originated in Central America and was domesticated by a Mesoamerican civilization. The historical biogeography of Caricaceae involved a long-distance dispersal event from Africa to the Neotropics during the Late Eocene. The deepest divergence in the Neotropics dates to the Oligocene-Miocene boundary and involves a split between a Central American and a (mostly) South American clade, suggesting range expansion across the Panamanian Isthmus. In the New World, diversification during the Miocene seems to be related to the main events of mountain building that formed new habitats and barriers, and to the climate cooling responsible by the expansion of dry habitats. The Pleistocene major climate change in Africa parsimoniously relates to the inferred divergence time of ancient West and East African populations.

The last part is dedicated to the evolution of chromosome numbers in the Caricaceae and includes counts for species from three genera (*Cylicomorpha, Horovitzia, Jarilla*) that have never been investigated before. Before my study, all published counts for Caricaceae were $2n = 18$, but preliminary results show that *Horovitzia cnidoscoloides* presents $2n = 16$, and two species of *Jarilla* (*J. caudata* and *J. heterophylla*) present $2n = 14$, indicating that chromosomal rearrangements resulting in the reduction of the chromosomes number may have occurred in the most recent common ancestor of this small clade.

Contents

I. General Introduction

Biological systematics aims to document and understand the history and diversity of life on Earth. Among other steps, this requires naming biological entities, which involves description, classification, and applying the rules of nomenclature. Before the molecular age, which began in the 1950s, systematics was based mostly on morphology. Today, however, much of systematics focuses on the phylogenetic relationships among species and higher clades, based on DNA sequences. While such "molecular systematics" has revolutionized our understanding of the evolution of organisms, it does not provide all the information required to name and describe the World's species in a recognizable and universal manner.

Concern about the loss of biodiversity is widespread around the Globe and affects the citizens of all countries, whether on Pacific Islands threatened by raising sea levels or in densely populated industrial regions of Europe. Understanding the extent and causes of biodiversity loss requires, among other fields of science, also systematic research. This is because biodiversity research provides the basis for proper species identification and the permanent preservation and documentation in public collections of examples of as many species and forms as possible. Efficient species identification is hampered by overly technical literature that is not updated and, often, expensive and inaccessible. Traditional taxonomic literature also usually fails to take full advantage of modern tools, such as electronic color images, much less the millions of named and unnamed images already available online and accessible for image recognition software (Kress 2004; Belhumeur et al. 2008; Shamir et al. 2010). Additionally, the high number of synonymous names (see below) contributes to the problem of finding the right name for an organism. In order to identify species properly and efficiently one needs identification keys, detailed species descriptions, and precise distribution maps. Most important is that this information must be accessible, easy to understand and be associated with illustrations. Identifying species correctly is pivotal for exploring chemical and economic properties of wild organisms, prioritizing areas for conservation, and assessing extinction risks. It can also help making the public aware of biodiversity and thereby help conservation.

1

To produce a meaningful picture of life on Earth it is necessary to bring together the knowledge available on each species (e.g., specimen-based geographic occurrence, morphological descriptions, revised nomenclature, chromosome numbers, reproductive biology, DNA sequences, etc.) and combine it with other types of information, such as climate and geological data. Today's Information Technology (IT)-infrastructure and bioinformatic tools set the stage for a new era of systematics in which the burden of taxonomic work is alleviated by ready access to public repositories with images of specimens, including types, and the literature of the past 260 years (back to Linneaus, 1753, the starting point of botanical nomenclature, and even older literature). Other molecular, geographical, ecological, and physical (geology and climate) data – often freely available – are also increasing exponentially.

For my doctoral research, I decided to use bioinformatic tools to produce an electronic monograph on a family of flowering plants, bringing together all available information and generating new data, including DNA sequences, chromosomes number, a completely revised nomenclature, identification keys, and well-illustrated morphological descriptions. I decided to focus on a suitably-sized family of flowering plants, the Caricaceae, which includes the economically important crop, *Carica papaya*. Caricaceae are well suited for applying and testing modern systematic tools because of (*i*) their economic importance (almost all species are used at least locally), which has resulted in numerous studies on chemistry, pharma-ceutical properties, and genomics that meant my study would be broadly useful, (*ii*) their relatively few species (34) but numerous available names (233), which has to do with the economic importance of papaya and the family's distribution in the humid tropics (still under-collected, with the scarcity of specimens [in all reproductive stages] contributing to unclear species boundaries), (*iii*) their relatively high number of red-listed species (six species listed in the IUCN, 2013), which meant it was important to better document species' geographic ranges, and (*iv*) the family not having been the focus of a recent monograph. The family's geographic occurrence in Africa, Central America, and South America also made it biogeographically interesting.

In the first part of this General Introduction, I expand on the importance of using bioinformatic tools to make the taxonomic effort more efficient and accessible to different communities of people. In the second part, I summarize basic knowledge on the Caricaceae and clarify the questions about the family's phylogeny, biogeography, and chromosome evolution that motivated my research.

The Problem of Synonyms and the Importance of Taxonomic Monographs

So far, there are c. 1.9 million accepted (named) species on Earth, from which only 66,307 represent microbial diversity (Chapman 2009). For flowering plants, there are 1,040,426 scientific names, 298,900 of them accepted, 263,925 unclear, and 477,601 (45.9%) synonyms (The Plant List 2010). A review of the problem of synonymous names shows that in plant groups that have been monographed, 58 to 78% of the published names turned out to be synonyms (Scotland and Wortley 2003). Such levels of synonymy are a serious problem. Moreover, the rate at which new synonyms are produced seems to be increasing linearly with the rate at which new species are being described (Wortley & Scotland 2004; Fig. 1). The high levels of synonymy and the scarcity of taxonomic treatments of larger groups are major impediments to the recognition of "good" species (because the increasing numbers of synonyms make it ever harder to study and sort the type material). This contributes to the relatively slow rate at which new species are being recognized as such and then described. Thousands of already collected new species await discovery in scientific collections (Bebber et al. 2010).

Besides slowing down systematic and evolutionary research, synonymous names also hamper the prediction of extinction rates, which requires the knowledge of how many species there are and what their range sizes are (Pimm et al. 1995). Synonyms usually result in too small species ranges and thus perhaps exaggerated estimates of endangered species because each name will be associated with its own "species" range. Lastly, synonymous names make it difficult to find published information on a particular biological entity, hampering the use of species for medical or any other kind of purpose, because users cannot know which names refer to which good species.

A taxonomic monograph brings together the information pertaining to all names that have ever been published for some group of organism and is the only way to assess, and reassess, the status of a name as either a synonym or a biological species. Monographer often also carry out some phylogenetic work based on DNA sequences from a representative subset of the specimens they have collected or loaned and then reach a conclusion about which names refer to which species, based on combined morphological, geographic, phylogenetic, and more rarely phylogeographic data. After defining species boundaries, the next step in monography is to summarize the key characteristics of the accepted species (i.e., describe the species), construct a key for their identification, and prepare an authoritative list of the accepted and synonymized names. Taxonomic revision (by monographs) is the only known mechanism for achieving quality control in taxonomy and for reducing the number of synonymous names that clutter up databases and hinder progress in our knowledge of the World's biodiversity and its conservation status. However, revisionary work produced by taxonomists (whether in floras or monographs) is of little utility if it is produced at a glacial pace and hard to access.

The idea that taxonomic research could be sped up by "moving into the electronic age" has been advocated for at least 10 years (Godfray 2002, 2007; Wilson 2003; Kress 2004; Scotland and Wood 2012). Advocates hold that taxonomic information creation, testing, and access can all benefit from what has been called "cyber-taxonomy" or "e-taxonomy" (Zauner 2009; Wheeler and Valdecasas 2010). Indeed, species descriptions are now increasingly being published online with cutting-edge publication technology that is improving automated linkage of different kinds of electronic information. The ultimate goal – which can now be achieved – is to summarize and disseminate the existing knowledge about the Earth's species and higher taxa (Blagoderov et al. 2010; Penev et al. 2010). However, different from species-level work, monography has not picked up its pace in spite of all tools available. That is why I chose to work on an e-monograph as part of my doctoral research.

Brassicales and the Caricaceae

The Caricaceae and their sister family Moringaceae are part of the mustard-oil plant clade or Brassicales, which also comprises 15 other families. Among them is the cabbage family (Brassicaceae), which includes *Arabidopsis thaliana*, the first plant to have its genome fully sequenced and a model organism for understanding plant biology, including developmental genetics, circadian rhythm, and many other aspects of plant life (The Arabidopsis Genome Initiative 2000; Müller and Grossniklaus 2010). *Carica papaya*, the main source of the World's papain, an enzyme widely used by food and pharmaceutical industries, was the 7th flowering plant selected for full genome sequencing (Ming et al. 2008). Since then, comparisons between papaya and *Arabidopsis* genome have improved our understanding of plant genome organization (e.g., Paterson et al. 2010). Because of the huge amount of genomic data available for these two species, Brassicales are now one of the most important plant groups for genome-wide studies.

The sister group relationships between Caricaceae and Moringaceae, and the position of both among the early-diverging Brassicales are well supported by molecular data (Beilstein et al. 2010). Moringaceae comprise 13 species in one genus, *Moringa*, and occur in seasonally dry regions of Namibia and Angola, the Horn of Africa, Madagascar, the Arabian Peninsula, Pakistan, and India (Olson 2002b). They are woody shrubs or trees often with swollen succulent trunks, and deciduous, 1- to 3-compound leaves that have conspicuous glands at the leaflet articulations (Olson 2002a, b; Fay and Christenhusz 2010). Their flowers resemble those of legumes, and their fruits are three-angled capsules. Especially striking is the growth form of Moringaceae, either that of bottle trees or tuberous shrubs, often with pachypodia, which are enlarged fleshy root or stem transitions (Olson and Rosell 2006). Some Caricaceae, such as *Jacaratia mexicana* and *J. corumbensis*, also are bottle-like trees or develop enormous tubers.

Distribution and Diversity of Caricaceae

After my taxonomic revision of all type specimens, Caricaceae consist of 34 species (and one formally named hybrid) in six genera, two of which occur in Africa and all others in the Neotropical region. The sole African genus has two species that are large trees with a gregarious habit and occur in humid, montane and submontane forests in East (*Cylicomorpha parviflora*) or West Africa (*C. solmsii*). *Vasconcellea*, the largest genus in the family, comprises 20 species plus a naturally occurring hybrid, *Vasconcellea* x *heilbornii* (Badillo 2000; Van Droogenbroeck et al. 2006). The genus has a center of species diversity in Northwestern South America, especially Ecuador, Colombia and Peru, with representatives in wet evergreen forests, seasonally tropical dry forests, and very arid regions.

The genus *Jacaratia*, with seven species of trees, is widespread in the lowlands of the Neotropics with only one species (*J. chocoensis*) occurring at altitudes up to 1,300 m in the Andes. *Horovitzia cnidoscoloides*, the only species in the genus, is a small tree reaching 6 m in height; it is known only from cloud forests of Sierra de Juaréz in Oaxaca, southern Mexico (Lorence and Colín 1988). *Jarilla* comprises three herbaceous species with perennial tubers that re-sprout annually during the wet season (Diaz-Luna and Lomeli-Sención 1992). *Carica papaya*, the only species in the genus *Carica*, is naturalized in the Neotropics, its northern range limit lies in Florida and the southern in Paraguay (Badillo 1971). However, truly wild papayas, which have much smaller fruits and thinner pulp than the cultivated ones, have only been found in the lowlands of Central America from Yucatan in Mexico, south to Belize and eastern Guatemala, and Costa Rica (Manshardt and Zee 1994; Coppens d'Eeckenbrugge et al. 2007).

Morphology, Pollination, Sexual Systems, and Chromosomes

My monographic research has clarified that most species of Caricaceae are trees or shrubs (three *Jarilla* species from Mexico and Guatemala are herbs, and *Vasconcellea horovitziana* is a liana). All species produce white or yellow latex from which the papain is extracted. Leaves vary from simple (entire to deeply lobed) to compound (palmate or trifoliolate).

The flowers in Caricaceae are unisexual, although bisexual flowers are found occasionally. Male flowers have nectaries on a small sterile ovary (called pistillode), while female flowers are devoid of nectar and also lack any stamen vestiges (Decraene and Smets 1999). Fruits are berries with many seeds that are surrounded by a mucilaginous aril; the testa can be ornamented or not. The basic morphological structure of flowers from both sexes is remarkably constant throughout the Caricaceae, with the few characters distinguishing species being found in the male flowers. The most useful taxonomic characters in the family are the shape of the anthers, the elongation (or not) of the connective, the seeds ornament, and fruit shape and color. For some species, shape of the inflorescence, flowers color, and leaf shape and venation are also distinctive characters.

Since female flowers produce neither nectar nor pollen and thus do not reward visiting insects, Baker (1976) introduced the term "mistake pollination" to describe pollination by foraging errors on the part of the moths that he observed visiting male and female papaya flowers in an orchard in Costa Rica. Bawa (1980) also suggested that the white and petaloid stigmatic lobes of the female flowers resemble the white corolla of male flowers of *Jacaratia dolichaula*, thereby increasing moth visits. (but without experimental evidence or data on actual visitor numbers to the male and female flowers). Although Bawa (1980) did not observed moths visiting female flowers he states that male flowers open and secrete nectar at dusk (between 5 and 6 p.m.), and that the pollen was deposited inside the narrow stigmatic canal of the female flowers, making Sphingidae (long-tongued, nocturnal moths) the best candidate as pollinators. The only other study mentioning pollination is by Aguirre et al. (2007) who observed moths and nocturnal bees (*Megalopta* sp.) visiting flowers of *Jacaratia mexicana*, but the bees were visiting mainly staminate flowers, while the moths visited both sexes (but also without experimental evidence or data on numbers of observed hours, flowers and visitors).

In Caricaceae, only one species (*Vasconcellea monoica*) has staminate and pistillate flowers on the same plant (monoecious); all other species have male and female flowers on different individuals and are thus dioecious. The only study of tree sex ratios in natural populations showed that in the dioecious *Jacaratia mexicana*, quite a few (up to 25%) of the male trees can have perfect flowers with functional ovaries – tested by rate of germination and seedling survival (Aguirre et al. 2007). The species can thus be called trioecious, meaning it has pure male trees, pure female trees, and some trees that function as males as well as females. Fruiting males have also been reported among cultivated plants of *Vasconcellea pubescens* and *Carica papaya* (Horovitz and Jiménez 1972).

Sex in *Carica papaya* is determined by sex chromosomes, morphologically identical to the autosomes. The first evidence of genetic sex determination in Caricaceae came from experiments that documented a 50:50 sex ratio among seedlings from female trees or hermaphrodite trees, i.e., the male trees with fertile female flowers (Storey 1953; Horovitz and Jimenez 1967). Genome sequencing has now confirmed that papaya has a small region that is recombination-suppressed and that is associated with maleness (Liu et al. 2004). The hermaphroditic trees have slightly different Y chromosomes, not the typical Y found in pure males, but instead a Yh chromosome (h stands for hermaphrodite; Liu et al. 2004; Ming et al. 2007; Wang et al. 2012). The male-specific region of the Y chromosome shares 98.8% sequence identity with the hermaphrodite-specific region of the Yh chromosome (Zhang et al. 2008). An XY genetic sex determination system is reported also for *Vasconcellea goudotiana*, *V. pubescens*, *V. parviflora*, and *V. pulchra* (Wu et al. 2010). So far, there is no report on Yh chromosomes for these species. Although there is now a large amount of genomic information on sex chromosomes of these few species, at the start of my research the chromosomes of only 11 of the 34 species from three of the six genera had been counted, all with $2n = 18$ (Heilborn 1921; Kumar and Srinivasan 1944; Bernardello et al. 1990; Caetano et al. 2008; Costa et al. 2008; Damasceno et al. 2009; Silva et al. 2012).

Taxonomic History and Previous Molecular Studies on Caricaceae

As expected, generic concepts in the Caricaceae have changed over the past 150 years as more material became available and especially with the advent of molecular data. Of the more than 230 available names in the family, 96 are basionyms, implying that slightly over half the names have been moved between genera. The first taxonomic treatment of the family was carried out by Alphonse De Candolle (1864) who dealt with the family under the name Papayaceae and recognized 22 species in three genera: *Papaya*, *Vasconcellea* divided in two sections (*Hemipapaya* and *Euvasconcellea*), and *Jacaratia*. Twenty-five years later Solms-Laubach (1889) accepted 28 species of Caricaceae in two genera, *Jacaratia*, and *Carica*. The latter with three sections: *Vasconcellea*, *Hemipapaya*, and *Eupapaya*. But it was only with the work of Victor Badillo (1971, 1993, 2000; Badillo et al. 2000) that the classification of the family gradually attained its current form.

Studies based on molecular data began in the 1990s (Jobin-Decor et al. 1997; Aradhya et al. 1999) and quickly revealed that species included in the section *Vasconcellea* of the genus *Carica* are more closely related to *Jacaratia* than to *Carica papaya* (the type species of section *Carica*). Reacting to these first molecular findings, Badillo (2000) reinstated *Vasconcellea* as a genus distinct from *Carica*, a decision supported by further molecular studies that found *Vasconcellea* and *Jacaratia* forming the sister clade to *C. papaya* (Van Droogenbroeck et al. 2002; Kyndt et al. 2005a; Chapter II). However, most species-level work has focused on the highland papayas, *Vasconcellea* (Van Droogenbroeck et al. 2004, 2006; Kyndt et al. 2005a,b), and no phylogenetic study prior to my own (Chapter II) included all species and genera of the Caricaceae. Therefore, when I began my research, phylogenetic relationships were still insufficiently understood.

The Main Geological and Climate Events Related to Caricaceae Biogeography

Among the most important historical events that shape current biodiversity are the uplift of mountains, climate cycles, and the isolation and reconnection of continents. The new climates, ecological gradients, and landscapes created by the combination of these processes set the stage for species evolution. Recent developments in the field of species distribution modeling, combined with phylogenetic approaches and geological data, is helping biologists to investigate the relative importance of ecological divergence *versus* geographic distance in the diversification of organisms (e.g., Graham et al. 2004; Loera et al. 2012). Adaptation to local environment combined with geographic isolation are the main forces driving speciation, certainly in montane areas with their highly variable landscapes and barriers formed during the processes of uplift.

Rainforests in East and West Africa are today isolated by an arid corridor acting as a dispersal barrier for rainforest taxa (Couvreur et al. 2008). Mountain building and climate oscillations that occurred since the Late Oligocene have promoted repeated expansion and retraction of these forests as documented by sedimentary records (Zachos et al. 2001; Trauth et al. 2009). Especially important were the alternating periods of aridity and humidity that characterized the climate in Africa during the late Cenozoic (Trauth et al. 2009). This dynamic vegetation history in Africa drove speciation and extinction in many groups of organism occurring today in the East and West African rainforests (e.g., Couvreur et al. 2008; Chatrou et al. 2009; Holstein & Renner 2011). For some groups, diversification seems to have been caused by adaptation to different habitats (Holstein and Renner 2011), while for others, vicariance (separation by intervening unsuitable habitat) was the primary factor for the formation of separate species (Couvreur et al. 2008).

In Mexico and Central America, too, complex orogenic events and historical climatic change were important factors driving the diversification of many groups (e.g., Loera et al. 2012; Bryson Jr and Riddle 2012). The main physiographic features of Mexico were formed during the late Cretaceous, but many geomorphological features developed gradually during the Miocene (Gómez-Tuena et al. 2007; Ferrari et al. 2012).

Especially interesting was the gradual formation of the Trans-Mexican Volcanic Belt (TMVD), which is a large mountain range that stretches from the Gulf of California in western Mexico to the Gulf of Mexico in the East. This belt was formed in several stages of volcanism from the West to the East with two main events during the Miocene, the first from 20 to 10 Mya, the second between 7.5 and 3 Mya (Gómez-Tuena et al. 2007; Ferrari et al. 2012). The Trans-Mexican Volcanic Belt created new geographical barriers between north and south, but also connected previously isolated montane biotas through the new east-western highland corridor (Anducho-Reyes et al. 2008; Bryson Jr and Riddle 2012).

South America was an isolated continent from the time of its separation from Africa until the closure of the Isthmus of Panama, which allowed a great biota exchange between North and South America. In addition, the closure of the Central America seaway caused extensive changes in the Atlantic Ocean circulation that may have promoted global climatic changes (Haug and Tiedemann 1998). Based on geochemical data and fossil records, the full closure of the isthmus, ending the communication between Caribbean and Pacific waters, occurred about 3 Mya (Coates et al. 2004; Woodburne 2010). Geologic evidence indicates that parts of the Isthmus emerged before the Miocene. For example, the San Blas range (a tectonic unit east of the Isthmus) was above sea level from the Late Eocene until the Miocene when it acted as a peninsula of North America (Farris et al. 2011; Montes et al. 2012). The major continental exchange of species, known as the Great American Biotic Interchange, however, is well dated to 3.1 to 2.5 Mya (Woodburne 2010; Gutiérrez-García and Vázquez-Domínguez 2013).

Another important geological event that shaped the Neotropical biota was the formation of the Andes, the largest mountain range in South America, and one of the most diverse in the world. The mountain building began during the Cretaceous, with the first events being the Southern and Central Andean uplift. Then the process continued in punctuated bursts, with the main episodes of uplift of the northern and central Andes occurring during the Miocene and Pliocene (Hoorn et al. 2010). Thus, the formation of the Andes affected different regions at different times, changing the climate and drainage patterns, as well as creating new habitats all over the continent. The huge impact of the Andean uplift on the diversification of many groups of organisms is clear (e.g., Hughes and Eastwood 2006; Antonelli et al. 2009; Chacón et al. 2012). Concomitant with the mountain building, during the Miocene, climatic change was promoting the expansion of dry habitats (i.e., dry forests, xerophytic shrublands, savannas and open grasslands) worldwide and also in Central and South America (Pound et al. 2011).

Research Questions and Aims

The main goals of my research were to test and help improve bioinformatic tools to increase and disseminate the taxonomic knowledge on the plant family Caricaceae (Chap. II); to generate new knowledge on the species of Caricaceae, reviewing the nomenclature and species boundaries (Chap. II); and to place all species in a morpho-ecological, geographical, and evolutionary context. To achieve these goals, I used phylogenetic and biogeographic approaches, investigated the evolutionary relationships among taxa based on plastid and nuclear DNA sequences, and studied the historical biogeography of the family in a molecular clock-dated framework. I also related the family's diversification to historical processes in Africa and in the Neotropics (Chap. IV). The questions I wanted to answer were: (*i*) what are the closest relatives of *Carica papaya*, (*ii*) when did the two African species diverge from each other and from their Neotropical relatives, and lastly (*iii*) how historical events relate to the diversification of Caricaceae. I also counted the chromosomes of four species (*Cylicomorpha parviflora, Jarilla heterophylla, J. caudata*, and *Horovitzia cnidoscoloides*) that had never been investigated before, partly because they had never been brought into cultivation before I collected seeds in Mexico and through contacts in Africa, which allowed me to grow these four species in the Botanical Garden of Munich (Chap. V).

II. Taxonomy in the Electronic Age: An e-Monograph of the Papaya Family (Caricaceae) as an Example [§]

Fernanda Antunes Carvalho[1, *], Denis Filer[2], Susanne S. Renner[1]

[1] Systematic Botany and Mycology, University of Munich (LMU), Menzinger Strasse 67, 80638 Munich, Germany
[2] Department of Plant Sciences, University of Oxford, South Parks Road Oxford, OX1 3RB, United Kingdom

[*] Corresponding author: antunesfc@gmail.com

[§] published in: *Cladistics*, 31(3), 321–329, June 2015, doi: 10.1111/cla.12095

Abstract

The need for taxonomists to take full advantage of biodiversity informatics has been clear for at least 10 years. Significant progress has been made in providing access to taxonomic resources online, including images of specimens, especially types; original species descriptions; and georeferenced collection data. However, in spite of persuasive calls for e-monography, there are few, if any, completed projects, even though monographic research is the only mechanism for reducing synonymous names, which are estimated to comprise 50% of all published names. Caricaceae is an economically important family of flowering plants from Africa and the Neotropics, best known for the fruit crop papaya. There is a large amount of information on the family, especially on chemistry, crop improvement, genomics, and the sex chromosomes of papaya, but up-to-date information on the 230 names and which species they might belong was not available. A dynamically updated e-monograph of the Caricaceae now brings together all information on this family, including keys, species descriptions, and specimen data relating the 230 names to 34 species and one hybrid. This may be the first taxonomic monograph of a plant family completely published online. The curated information will be continuously updated to improve the monograph's comprehensiveness and utility.

Introduction

The Plant List (2010) shows 1,040,426 published names for plants of which 29% are accepted, 25% of unclear status, and 46% considered synonymous with other species names. The problem of synonymous names arises because taxonomists inadvertently name the same species several times, usually because it is widespread and has been collected in far-apart regions and/or because widespread species often are morphologically variable, sometimes in correlation with their environment, making it difficult to assess species status until a dense collection series can be studied. In the flowering plants, there may be 3–4 synonyms for every accepted name (Scotland and Wortley, 2003;

Wortley and Scotland, 2004; Paton et al., 2008; The Plant List, 2010). The problem of synonymous names is by no means restricted to plants, although reliable estimates for all eukaryotes are difficult to obtain (Alroy, 2002; Mora et al., 2011). Synonymous names are not a harmless nuisance, and their rate seems to be increasing apace with the rate at which new species are described (Fig. 1). When it comes to conserving species or using species for medical or any other kind of purpose, synonymous names will result in two kinds of errors: they result in wrong, usually narrower, species range estimates than warranted because each name will be associated with its own "species" range; and they make it difficult to find material of, or published information on, a particular biological entity because users cannot know which names refer to which good species.

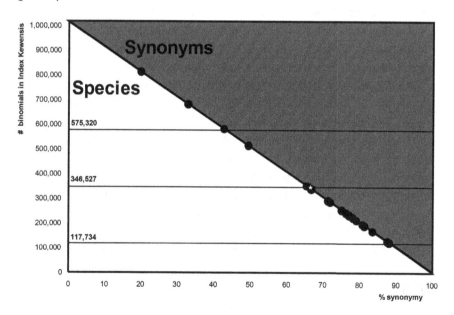

Fig. 1. Relationship between synonymy percentage and number of species from Wortley and Scotland (2004), reproduced with permission of the authors.

The assessment, and reassessment, of the status of a name as either a synonym or a good species is done during monographic research. Monographic research is based on bringing together the information pertaining to all names that have ever been published for some group, typically a genus or a family. This will include the publication in which a name was first proposed (the so-called protologue), all specimens to which the name has been applied (rightly or wrongly), the accepted names and their synonyms, morphological descriptions for each species, geographic coordinates of relevant collections, chromosome numbers, chemical traits, flowering or fruiting times, and DNA sequences from specimens given one or several of the names in question. A monographer will study the specimens, often do some phylogenetic work based on DNA sequences of a representative subset, and reach a conclusion about which names refer to which species. He/she next constructs a key to identify the accepted species and prepares an authoritative list of the accepted and synonymized names. Monography is the only known mechanism for achieving quality control in taxonomy and for reducing the number of synonymous names that clutter up databases and hinder progress in our knowledge of the World's biodiversity and its conservation status.

Because taxonomy is the portal to the entire information available about species, the need for taxonomic research to "move into the electronic age" has long been clear (Bisby et al., 2002; Godfray, 2002; Wilson, 2003; Kress, 2004; Wheeler, 2004; Scotland and Wood, 2012). Indeed species descriptions of animals and plants are now increasingly being published online (Blagoderov et al., 2010; Knapp, 2010; Penev et al., 2010; Knapp et al., 2011). Monography, however, has not followed suit, in spite of the availability of massive online databases of literature and digitized specimen, wikis, ever cheaper digital photography and microscopy (essential to the study of herbarium specimens), and dedicated platforms for taxonomy, such as the Botanical Research and Herbarium Management System (BRAHMS, http://herbaria.plants.ox.ac.uk/bol/) and Scratchpads (http://scratchpads.eu/). The new "cyber-taxonomy" or "e-taxonomy" (Zauner, 2009; Wheeler and Valdecasas, 2010) is reality only for species descriptions and lists of names but not yet for monographic research (Scotland and Wood, 2012). Although there are several ongoing taxon-centered initiatives (Appendix 1), to our knowledge no revision or monograph of any large group has been completed. The advantages of online monography, such as the possibility of including near-unlimited color images and the option of up-dating information, have thus not been realized.

Overview of the Electronic Monograph of Caricaceae and its Underlying Database

Here we present a recently completed electronic monograph of a plant family (Caricaceae), the result of research that brought together the available collections with digital libraries, digitized specimen data, and other taxonomic and methodological tools available, including DNA sequencing for barcoding the recognized species (Carvalho and Renner, 2012, 2013).

Caricaceae is a small family of flowering plants from Africa and the Neotropics, best known for the fruit crop *Carica papaya*. The family's economic importance lies not only in the papaya fruit, but also in the production of papain, a cysteine proteinase widely used in food and pharmaceutical industries. A search for the topics 'papaya' and 'papain' in Web of Knowledge retrieves approximately 20,823 and 42,100 citations, respectively (ReutersISI, 2013). Several Caricaceae are considered as unexploited crops because of their nutritive fruits, high concentration of papain-like enzymes, and resistance to pathogens (Kyndt et al., 2007; Ramos-Martínez et al., 2012). Among these are the so-called highland papayas, species of *Vasconcellea,* a genus thought to be synonymous with *Carica* until Badillo (2000) cleared up their morphological distinctness (Badillo, 2000). Molecular data have revealed that the closest relative of papaya is a clade of four species in Mexico and Guatemala entirely neglected by ecologists and breeders (Carvalho and Renner, 2012). The lack of knowledge before 2012 on the true closest relatives of papaya resulted in the assumption that the highland papayas (*Vasconcellea* species) were the best group to use in papaya improvement (Scheldeman et al., 2011; Coppens d'Eeckenbrugge et al., 2014).

As required in a taxonomic monograph, the e-monograph of Caricaceae (http://herbaria.plants.ox.ac.uk/bol/caricaceae) allocates all names (here 230) to recognized species (here 34 and one hybrid), providing a comprehensive data infrastructure for scientists and nonscientists alike. The database is being developed, managed and published online using BRAHMS (http://herbaria.plants.ox.ac.uk/bol) developed at the University of Oxford. In carrying out this research on the Caricaceae, we added a range of new features to BRAHMS that facilitate cyber-monography emphasizing thus the importance of close collaborations among taxonomists and bioinformaticians (Stein, 2008).

The e-monograph of Caricaceae and its underlying database, store (and make available) data and images on collections, herbarium specimens, literature, and the revised nomenclature (including accepted names, synonyms, *nomina nuda*, illegitimate names, and excluded names). The monographic research resulted in updated circumscriptions of the recognized species, including detailed plates (Fig. 2), and precise geographic distribution of all relevant collections. Links to supportive literature and high-resolution images of type specimens are provided for each species as are cross-references to databases, such as The Plant List, TROPICOS, IPNI, and GBIF. General information on the family, including its ecology, sex chromosomes, and molecular phylogeny is provided, along with identification keys to all genera and species.

All these data are accessible through BRAHMS online and summarized in Table 1. Searches by taxon, collector, geographic place name, and map area (Fig. 3) generate tables that can also be shown in text format. Images can be grouped and filtered, and viewed at different resolutions. Maps are available using clustered Google Maps or Google Earth, both configurable with zoom features. A detailed description of the methods used to build the e-monograph is given in Appendix 2.

Discussion

Among the challenges for taxonomy today are to incorporate results and insights from molecular phylogenetic work and to tackle the problem of the 46–50% synonymous names already published (Scotland and Wortley, 2003;

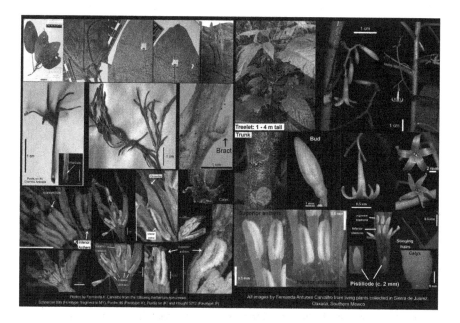

Fig. 2. Examples of species plates used to describe species in the website. To the left are images of details of male flowers and inflorescences based on herbarium specimens of *Vasconcellea longiflora*; to the right, images of living material of *Horovitzia cnidoscoloides*, one out of the four little known closest relatives of papaya.

Wortley and Scotland, 2004; The Plant List, 2010). Both challenges can only be addressed through monographic work in which species and genus circumscriptions are vetted and updated, based on the study of specimens and consideration of relevant phylogenetic results on relationships.

Reliably circumscribed and named species are also required to fulfill the promise of DNA barcodes, at least if that promise is finding names for unidentified specimens via matching of short DNA sequences (obviously, one can also match unnamed material to unnamed sequences). Simply increasing the rate of species discovery, while important, does not address either of these challenges because naming a newly discovered species does not require a complete assessment of all existing names that might apply (which would often take too much time). It is therefore likely that as the number of species descriptions increases (Costello et al., 2013), so does the number of newly created synonyms (Fig. 1).

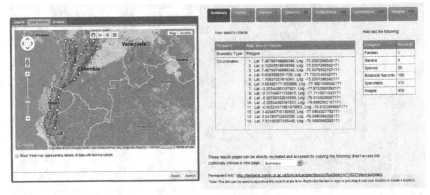

Fig. 3. Map search in BRAHMS. The left figure shows a polygon that can be drawn by the user to delimit the area of interest, in this case, the Andes from northern Peru to northern Colombia. To the left is a summary of the results, which includes number of genera, collections, specimens, and images available in the database. It also provides the coordinates of the polygons, which can be used to create a shape file.

A well-resolved, expert-vetted nomenclature and detailed information on the distribution of species are of great importance for many fields of research (Yesson et al., 2007; Bortolus, 2008; Patterson et al., 2010; Lis and Lis, 2011; Santos and Branco, 2012). However, high-quality data produced by taxonomists in revisions and monographs are of little use unless widely accessible (Kress, 2004). This is especially important for economically important groups, which often are also groups with a high rate of nomenclatural changes (as is the case for Caricaceae). Open-access information to this highly organized set of online data and images for the Caricaceae benefits the scientific community broadly as well as those working on the food and medicinal aspects of the family. This includes the community of herbarium curators, researchers focusing on papaya genomics (Fig. 4A), breeders, and the non-scientific public. In addition, georeferenced specimens are the basis for the growing field of bioclimatic modeling (Fig. 4B) and for a reliable baseline to document the effects of ongoing climatic changes on plant ranges.

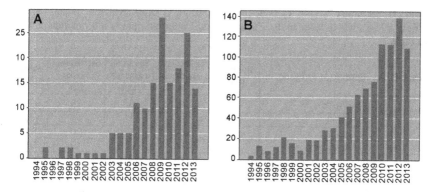

Fig. 4. (A) Number of published studies with the topic search fields "Caricaceae" and "genome"; a total of 168 records were found. **(B)** Number of published studies on Bioclimatic Modeling per year; in total 1,002 records. (Web of Knowledge accessed 18 November 2013).

In the case of the papaya family, the most recent taxonomic accounts were by Victor Manuel Badillo (1920–2008; http://herbaria.plants.ox.ac.uk/bol/caricaceae#badillo) a Venezuelan taxonomist who dealt with c. 200 names described in the family, 64 of these basionyms (meaning that the remainder result from changing generic concepts). The work of Badillo (1971, 1993, 2000) is poorly accessible, and since his last publication 13 years ago (Badillo, 2001) no further taxonomic work on the Caricaceae has been published. Meanwhile, molecular work on the family took off (Van Droogenbroeck et al., 2002; Kyndt, Romeijn-Peeters, et al., 2005; Kyndt, Van Droogenbroeck, et al., 2005; Carvalho and Renner, 2012). The IUCN Red List of Threatened Species (IUCN, 2013) lists six endangered species of Caricaceae, none under the correct name; the new e-monograph available at http://herbaria.plants.ox.ac.uk/bol/caricaceae, now includes updated information on the vulnerability of species that together with the geographic and ecological information should help in conservation efforts.

Table 1. Summary of the Caricaceae e-monograph data available online as of 13 Feb. 2014. Invalid, Illegitimate, Excluded and Uncertain names are not included in this table

Genera (6)	Species (34 + 1 hybrid)	Synonyms (160)	Collections examined (2950)	Specimens examined (4337)	Georeferenced collections (2204)	Images (10988)
Cylicomorpha	C. parviflora Urb.	1	36	57	28	246
	C. solmsii Urb.	1	18	27	12	158
Carica	C. papaya L.	21	590	773	30	1911
Horovitzia	Horovitzia cnidoscoloides	1	97	26	19	68
Jarilla	Jarilla chocola Standl.	1	37	52	36	136
	Jarilla caudata (Brandegee) Standl.	4	50	62	48	159
	Jarilla heterophylla (Cerv. ex La Llave) Rusby	4	71	85	69	219
Jacaratia (7 species)	J. digitata (Poepp. & Endl.) Solms-Laub.	3	178	251	167	512
	J. spinosa (Aubl.) A.DC.	8	209	329	190	849
	J. chocoensis A.H.Gentry & Forero	0	15	21	15	31
	J. corumbensis Kuntze	3	41	86	34	281
	J. dolichaula (Donn.Sm.) Woodson	1	128	172	120	450
	J. mexicana A. DC.	7	142	158	132	500
	J. heptaphylla (Vell.) A.DC.	1	30	37	26	134
Vasconcellea (21 species and 1 hybrid)	V. candicans (A.Gray) A.DC.	3	26	38	19	141
	V. cauliflora (Jacq.) A.DC.	8	111	159	87	452
	V. crassipetala (V.M.Badillo) V.M.Badillo	1	6	16	5	55
	V. glandulosa A.DC	9	80	153	71	419
	V. goudotiana Triana & Planch.	4	20	35	11	107
	V. horovitziana (V.M.Badillo) V.M.Badillo	1	13	34	3	125
	V. longiflora (V.M.Badillo) V.M.Badillo	1	6	10	2	26
	V. microcarpa (Jacq.) A.DC.	22	401	774	336	1508
	V. monoica (Desf.) A.DC.	7	32	70	14	156
	V. omnilingua (V.M.Badillo) V.M.Badillo	1	2	3	1	16
Vasconcellea (21 species and 1 hybrid)	V. palandensis (V.M.Badillo, Van den Eynden & Van Damme) V.M.Badillo	1	3	6	3	27
	V. parviflora A.DC.	5	61	109	46	323
	V. pubescens A.DC.	10	102	230	68	501
	V. pulchra (V.M.Badillo) V.M.Badillo	1	13	36	10	98
	V. quercifolia A.St.-Hil.	14	157	253	114	675

Genera (6)	Species (34 + 1 hybrid)	Synonyms (160)	Collections examined (2950)	Specimens examined (4337)	Georeferenced collections (2204)	Images (10988)
continued	*V. sphaerocarpa* (García-Barr. & Hern.Cam.) V.M.Badillo	1	13	25	12	60
	V. sprucei (V.M.Badillo) V.M.Badillo	1	27	57	7	184
	V. stipulata (V.M.Badillo) V.M.Badillo	1	19	40	15	115
	V. weberbaueri (Harms) V.M.Badillo	1	8	31	7	82
	V. chilensis Planch. ex A.DC.	3	16	38	6	86
	V. x heilbornii (V.M.Badillo) V.M.Badillo	9	31	84	13	188

A major problem in building the Caricaceae database was to gather data from different herbaria that use different standards and field definitions. This occurs despite proposals to standardize biodiversity databases such as HISPID (Herbarium Information Standards and Protocols for Interchange of Data; Conn, 1995), Darwin Core (Wieczorek et al., 2012) and ABCD (TDWG, 2013). The last two recommended by the Taxonomic Database Working Group (TDWG). New software and platforms are being developed each year by different institutions, but the communication among them is not being improved at the same pace. Database mapping can be used to integrate two distinct data models. However, if the same piece of information is digitized slightly different among institutes, queries that address multiple databases may not be adequately solved (Willemse et al., 2008). Standardization in data entry would increase the value of freely available biodiversity data by facilitating the use and re-use, distribution, and integration of this information. Initiatives, like speciesLink (http://splink.cria.org.br/), which integrates primary data from biological collections deposited in different scientific collections using Darwin Core standards, are laudable and should be linked to worldwide programs, such as the Encyclopedia of Life (EOL, http://eol.org/) and the Global Biodiversity Information Facility (GBIF, http://www.gbif.org/).

With the development of digital photography technology, professional and amateur photographers are unknowingly discovering and informally documenting new species by placing images of plants and animals in online image databases (Winterton et al., 2012). Species identification via images is becoming more and more important, and freely available e-monographs that combine images (which will be picked up, for example, by "google images") with professionally curated names and descriptions can support such citizen science. Systematists, however, are not yet producing freely accessible taxonomic monographs (or floras) despite more than ten years' worth of admonitions (Bisby et al., 2002; Godfray, 2002; Wilson, 2003; Kress, 2004; Wheeler et al., 2004; Scotland and Wood, 2012). This probably has two (related) reasons: the small number of people in a position to populate the existing cyber-infrastructure with data and the pressure for publishing in citable journals or monograph series. Overcoming the second problem will require citation of online publications as has long been standard in physics, mathematics, computer science, and chemistry.

The e-monograph of Caricaceae includes all features of a traditional monograph (Marhold et al., 2013), and is a single portal to access information on all taxon names, thus facilitating the communication among different groups of researchers. Different from any hard-copy work, however, it is rapidly searchable and links specimens and species to other kinds of data and knowledge; for example, specimens used in DNA isolation are linked to the respective GenBank entries. Another obvious advantage of online monography is the ease of updating. A newly discovered species, a range expansion, or a newly available set of images can easily be added to an online database, but not to a printed monograph. E-monographs will greatly improve access to knowledge about species, while at the same time feeding other databases with invaluable information for scientific research, society, and industry. As John Kress (2004, p. 2152 and 2127) envisioned, "With remote wireless communication the field botanist will be able to immediately compare newly collected plants with type specimens and reference collections archived and digitized in museums thousands of miles away. ...[The e-monographs] of the future, including web-based, computer-based, image-based, and even DNA-based products, are ... fulfilling new functions that paper-based and word-based floras of the past could never attain."

Acknowledgments

We thank all herbaria that are providing open access to images of specimens, indispensable for the development of e-taxonomy; all curators of the herbaria visited by FAC during the development of the monograph; Andrew Liddell at Plant Sciences, Oxford, for his work on the BRAHMS online system; Carmen Benítez for useful information on Victor Badillo and for giving FAC access to all his literature, including original hand-writings; Theodor C. H. Cole is providing editorial support in reviewing the e-monograph.

Appendix 1. Weblinks and short descriptions of e-taxonomy projects.
Only websites aiming to be a taxonomic monograph and already containing considerable information are included. All weblinks were accessed on 6 April 2014.

Name of the website and weblink	Short-description
Hymenopteran systematics http://hymenoptera.ucr.edu/mx-database	Detailed information on the main groups of Hymenoptera and a dichotomous key for identification of North American eulophid genera. Missing distribution data and images of specimens.
EuphORBia http://www.euphorbiaceae.org/pages/data_portal.html	General information on the genus *Euphorbia*. Missing identification keys, distribution maps and descriptions of the species. Also, there is no information or images of specimens and the project seems not to have advanced since 2010.
Milichiidae online http://milichiidae.info/content/milichiella-online-revision	Revision of Milichiidae or freeloader flies including descriptions, dichotomous keys, list of specimens and distribution maps for most of the taxa.
Bombus Bumblebees of the world http://www.nhm.ac.uk/research-curation/research/projects/bombus/	A taxonomic review of the group, including an overview of all world bumblebees species with interactive keys, distribution maps and descriptions for most of the taxa.
Lacistemataceae http://tolweb.org/Lacistemataceae	A short overview of the plant family Lacistemataceae including nomenclature for all taxa. Detailed species descriptions with images, information on vouchers and distribution maps are missing for most of the taxa.
Solanaceae Source. A global taxonomic resource for the nightshade family http://solanaceaesource.org/	An introduction, including general morphology, distribution, and other data. Also information on taxonomic status of genera and species. Distribution maps and images of many taxa are missing, and the monograph so far covers a small percentage of the c. 4000 species.

25

FishBase http://www.fishbase.org/	A database of world's fishes with information on common and scientific names, descriptions and environmental requirements of each taxon, images and distribution maps.
WoRMS World Register of Marine Species http://www.marinespecies.org/	Detailed information on nomenclature (95% of accepted names in the database were already checked). Missing descriptions and illustration of many taxa.
Miconieae http://sweetgum.nybg.org/melastomataceae/	Descriptions, phenology, distribution and nomenclature of many taxa from the tribe Miconieae, which includes c. 1800 species. Identification keys and descriptive images of each accepted name are missing, although images of specimens are available
The genus *Leucaena* http://herbaria.plants.ox.ac.uk/bol/leucaena	Monograph of the genus *Leucaena* with two dichotomous keys (using vegetative, floral or fruit characters), detailed species descriptions, nomenclature, images of each accepted species and specimen list. Missing images of specimens.
ILDIS International Legume Database & Information Service http://www.ildis.org/	Taxonomic database of Fabaceae with list of taxa and taxon status. Includes short descriptions for the accepted names but lacks images.
Xper^2 http://lis-upmc.snv.jussieu.fr/xper2/infosXper2Bases/en/liste-bases.php	List of knowledge bases using Xper^2, i.e., a list of interactive keys for different taxonomic groups. The keys constitute morphological databases, many of them well-illustrated. No information on the taxonomic status of the names is provided, and there are no images of specimens.
Scratchpads - Biodiversity online http://scratchpads.eu/explore/sites-list	List of websites produced with scratchpads. Although the number of available websites is large, few are active projects with much information, while many are rather incomplete and appear stagnant.
BRAHMS online websites http://herbaria.plants.ox.ac.uk/bol/brahms/Websites	List of websites published using BRAHMS. Most are databases with name list and sometimes information on taxonomic status.

Appendix 2. Methods used to produce the e-monograph of Caricaceae

The database is developed, managed and published online using the Botanical Research and Herbarium Management System (BRAHMS, http://herbaria.plants.ox.ac.uk/bol) developed at the University of Oxford. The principal reasons for choosing this software were (i) its established use in more than 60 countries around the world, facilitating communication among databases and researchers at different institutions, (ii) a user-friendly interface with many tutorials, (iii) freely available resources, and (iv) presence of a powerful module for publishing data online.

We imported into BRAHMS draft lists of taxon names available in the International Plant Names Index (http://www.ipni.org/) and TROPICOS (http://www.tropicos.org/). Duplicated names were marked and selected for deletion using the BRAHMS editing function "Tag identical entries ". Protologues were then located on the web and linked to each name in the database. Main providers of old relevant literature at this stage were Botanicus (http://www.botanicus.org/) and BHL (http://www.biodiversitylibrary.org/) both dynamically accessed using BRAHMS web toolbar links. Smaller online libraries, such as Internet Archive (http://archive.org/) and Gallica (http://gallica.bnf.fr/), were also important for texts not found elsewhere. Protologues not available online were acquired using the library facilities of the University of Munich. Following the entry of the protologues,

Details of type specimens were initially entered in the database following the most recent taxonomic work on Caricaceae (Badillo, 1971, 1993) and updated later on. Data relevant to nomenclature and taxonomic decisions, such as synonymization, the taxonomic status of each name and legitimacy, were edited using further formatting tools in BRAHMS. We kept in the database also not validly published names (*nomina nuda*) because some still populate other digital databases.

Web links to high-resolution images (as provided by some herbaria) were then added to the specimen records. Herbarium specimens form the base of this e-monograph, and the first author photographed all specimens she could find in relevant herbaria of Latin America (BHCB, GUADA, HUEFS, IBUG, MBM, MEXU, MY, R, RB, UPCB, VEN), North America (NY, F), and Europe (B, BM, GB, K, M, OXF, P, W, WU, GENT, S), either in loaned material or during personal visits. CGE, FI, INPA, IAN, and MG provided images of important Caricaceae specimens. At least two images of each specimen were taken: first, the label (to facilitate data entering) and second, the complete specimen. Images of details, such as stipules, trichomes and venation, were also taken for some specimens. All images were processed and renamed using tools provided in BRAHMS, following the tutorial available on http://herbaria.plants.ox.ac.uk/bol/caricaceae. Photography relied on the Macro function of a camera RICOH CX5 at resolutions of 3, 5, 7, or 10 megapixels and the digital microscope Dino-Lite AM-413ZT, a portable device hooked up to a laptop. Measurements of morphological characters were made with either ImageJ (http://rsbweb.nih.gov/ij/) or DinoCapture 2.0 (http://www.dino-lite.com/support.php), and a morphological database was built with Xper^2 (Ung et al., 2010), which allows the creation of interactive keys and can be integrated with the BRAHMS online system. Using one of the text reporting tools in BRAHMS, the Xper^2 database was exported to text format to generate standardized species descriptions.

For distribution maps, coordinates were taken from the specimen labels when available and then checked on Google Earth, easily accessible through a BRAHMS toolbar. Localities of collections without coordinates were found using available gazetteers and then corrected using Google Earth, following guidelines provided by Garcia-Milagros and Funk (2010). We also used other information present on specimen labels, such as elevation, distance from other locations (e.g., "10 km South of..."), and habitat (e.g., "across the river, up the slopes, in a dry area"). Where locality names were not in Google Earth, we checked historical maps, Wikipedia, and studies of botanical itineraries. The sources of all geographic coordinates were added to the field "llorig" (source of the latitude and longitude values) in BRAHMS.

References

Alroy, J. 2002. How many named species are valid? Proc. Natl. Acad. Sci. U.S.A., 99, 3706–11.
Badillo, V. M. 1971. Monografia de la familia Caricaceae. Asociación de profesores, Universidad Central de Venezuela, Maracay.
Badillo, V. M. 1993. Caricaceae. Segundo Esquema. Rev. Fac. Agron. (UCV), 43, 1–111.
Badillo, V. M. 2000. *Vasconcella* St.-Hil. (Caricaceae) con la rehabilitación de este último. Ernstia, 10, 74–79.
Badillo, V. M. 2001. Nota correctiva *Vasconcellea* St. Hill. y no *Vasconcella* (Caricaceae). Ernstia, 11, 75–76.
Bisby, F. A., Shimura, J., Ruggiero, M., Edwards, J., Haeuser, C. 2002. Taxonomy, at the click of a mouse. Nature, 418, 367.
Blagoderov, V., Brake, I., Georgiev, T., Penev, L., Roberts, D., Ryrcroft, S., Scott, B., Agosti, D., Catapano, T., Smith, V. S. 2010. Streamlining taxonomic publication: a working example with Scratchpads and ZooKeys. Zookeys, 28, 17–28.

Bortolus, A. 2008. Error cascades in the biological sciences: the unwanted consequences of using bad taxonomy in ecology. Ambio, 37, 114–8.

Carvalho, F. A., Renner, S. 2013. Correct names for some of the closest relatives of *Carica papaya*: A review of the Mexican/Guatemalan genera *Jarilla* and *Horovitzia*. PhytoKeys, 29, 63–74.

Carvalho, F. A., Renner, S. S. 2012. A dated phylogeny of the papaya family (Caricaceae) reveals the crop's closest relatives and the family's biogeographic history. Mol. Phylogenet. Evol., 65, 46–53.

Conn, B. J. 1995. HISPID - Herbarium Information Standards and Protocols for Interchange of Data. Version 3. Retrieved December 13, 2013, from http://plantnet.rbgsyd.nsw.gov.au/HISCOM/HISPID/HISPID3/H3.html

Coppens d'Eeckenbrugge, G., Drew, R., Kyndt, T., Scheldeman, X. 2014. *Vasconcellea* for Papaya Improvement. In: Ming, R., Moore, P. H. (Eds.), Genetics and Genomics of Papaya (1st ed.). Springer, New York, NY, pp. 47–79.

Costello, M. J., May, R. M., Stork, N. E. 2013. Can we name Earth's species before they go extinct? Science, 339, 413–416.

Garcia-Milagros, E., Funk, V. A. 2010. Improving the use of information from museum specimens: Using Google Earth to georeference Guiana Shield specimens in the US National Herbarium. Front. Biogeogr., 2.3, 71–77.

Godfray, H. C. J. 2002. Challenges for taxonomy. Nature, 417, 17–19.

IUCN. 2013. The IUCN Red List of Threatened Species. Version 2013.2. www.iucnredlist.org. Retrieved December 08, 2013, from www.iucnredlist.org

Knapp, S. 2010. Four new vining species of *Solanum* (Dulcamaroid Clade) from montane habitats in tropical America. PLoS One, 5, 1–8, e10502.

Knapp, S., McNeill, J., Turland, N. J. 2011. Changes to publication requirements made at the XVIII International Botanical Congress in Melbourne - what does e-publication mean for you? Taxon, 60, 250.

Kress, W. J. 2004. Paper floras: how long will they last? A review of Flowering Plants of the Neotropics. Am. J. Bot., 91, 2124–2127.

Kyndt, T., Romeijn-Peeters, E., Van Droogenbroeck, B., Romero-motochi, J. P., Gheysen, G., Goetghebeur, P. 2005. Species relationships in the genus *Vasconcellea* (Caricaceae) based on molecular and morphological evidence. Am. J. Bot., 92, 1033–1044.

Kyndt, T., Van Damme, E. J. M., Van Beeumen, J., Gheysen, G. 2007. Purification and characterization of the cysteine proteinases in the latex of *Vasconcellea* spp. FEBS J., 274, 451–62.

Kyndt, T., Van Droogenbroeck, B., Romeijn-Peeters, E., Romero-Motochi, J. P., Scheldeman, X., Goetghebeur, P., Van Damme, P., Gheysen, G. 2005. Molecular phylogeny and evolution of Caricaceae based on rDNA internal transcribed spacers and chloroplast sequence data. Mol. Phylogenet. Evol., 37, 442–59.

Lis, J. A., Lis, B. 2011. Is accurate taxon identification important for molecular studies? Several cases of faux pas in pentatomoid bugs (Hemiptera: Heteroptera: Pentatomoidea). Zootaxa, 2932, 47–50.

Marhold, K., Stuessy, T., Agababian, M., Agosti, D., Alford, M. H., Crespo, A., Crisci, J. V, Dorr, L. J., Ferencová, Z., Frodin, D., Geltman, D. V, Kilian, N., Linder, H. P., Lohmann, L. G., Oberprieler, C., Penev, L., Smith, G. F., Thomas, W., Tulig, M., Turland, N., Zhang, X.-C. 2013. The Future of Botanical Monography: Report from an international workshop, 12–16 March 2012, Smolenice, Slovak Republic. Taxon, 62, 4–20.

Mora, C., Tittensor, D. P., Adl, S., Simpson, A. G. B., Worm, B. 2011. How many species are there on Earth and in the ocean? PLoS Biol., 9, e1001127.

Paton, A. J., Brummitt, N., Govaerts, R., Harman, K., Hinchcliffe, S., Allkin, B., Lughadha, E. N. 2008. Towards Target 1 of the Global strategy for plant conservation: a working list of all known plant species — progress and prospects. Taxon, 57, 602–611.

Patterson, D. J., Cooper, J., Kirk, P. M., Pyle, R. L., Remsen, D. P. 2010. Names are key to the big new biology. Trends Ecol. Evol., 25, 686–91.

Penev, L., Kress, W. J., Knapp, S., Li, D.-Z., Renner, S. 2010. Fast, linked, and open – the future of taxonomic publishing for plants: launching the journal PhytoKeys. PhytoKeys, 14, 1–14.

Ramos-Martínez, E. M., Herrera-Ramírez, A. C., Badillo-Corona, J. A., Garibay-Orijel, C., González-Rábade, N., Oliver-Salvador, M. D. C. 2012. Isolation of cDNA from *Jacaratia mexicana* encoding a mexicain-like cysteine protease gene. Gene, 502, 60–8.

ReutersISI, T. 2013. Web of Knowledge. Retrieved October 19, 2013, from http://apps.webofknowledge.com/

Santos, A. M., Branco, M. 2012. The quality of name-based species records in databases. Trends Ecol. Evol., 27, 6–7.

Scheldeman, X., Kyndt, T., Coppens d'Eeckenbrugge, G., Ming, R., Drew, R., Van Droogenbroeck, B. V., Van Damme, P., Moore, P. H. 2011. *Vasconcellea.* In: Kole, C. (Ed.), Wild Crop Relatives: Genomic and Breeding Resources. Tropical and Subtropical Fruits. Springer-Verlag, Berlin Heidelberg, pp. 213–249.

Scotland, R. W., Wood, J. R. I. 2012. Accelerating the pace of taxonomy. Trends Ecol. Evol., 27, 415–6.

Scotland, R. W., Wortley, A. H. 2003. How many species of seed plants are there? Taxon, 52, 101–104.

Stein, L. D. 2008. Towards a cyberinfrastructure for the biological sciences: progress, visions and challenges. Nat. Rev. Genet., 9, 678–88.

TDWG. 2013. Biodiversity information standards. Retrieved November 16, 2013, from http://rs.tdwg.org/dwc/

The Plant List. 2010. Version 1. Retrieved December 10, 2013, from http://www.theplantlist.org/

Ung, V., Dubus, G., Zaragüeta-Bagils, R., Vignes-Lebbe, R. 2010. Xper2: introducing e-taxonomy. Bioinformatics, 26, 703–4.

Van Droogenbroeck, B., Breyne, P., Goetghebeur, P., Romeijn-Peeters, E., Kyndt, T., Gheysen, G. 2002. AFLP analysis of genetic relationships among papaya and its wild relatives (Caricaceae) from Ecuador. Theor. Appl. Genet., 105, 289–297.

Wheeler, Q. D. 2004. Taxonomic triage and the poverty of phylogeny. Philos. Trans. R. Soc. Lond. B. Biol. Sci., 359, 571–83.

Wheeler, Q. D., Raven, P. H., Wilson, E. O. 2004. Taxonomy: impediment or expedient? Science, 303, 285.

Wheeler, Q. D., Valdecasas, A. G. 2010. Cybertaxonomy and ecology. Nat. Educ. Knowl., 1, 6.

Wieczorek, J., Bloom, D., Guralnick, R., Blum, S., Döring, M., Giovanni, R., Robertson, T., Vieglais, D. 2012. Darwin Core: an evolving community-developed biodiversity data standard. PLoS One, 7, e29715.

Willemse, L. P. M., Welzen, P. C. Van, Mols, J. B., Taxon, S., May, N. 2008. Standardisation in data-entry across databases: avoiding babylonian confusion. Taxon, 57, 343–345.

Wilson, E. O. 2003. The encyclopedia of life. Trends Ecol. Evol., 18, 77–80.

Winterton, S. L., Guek, H. P., Brooks, S. J. 2012. A charismatic new species of green lacewing discovered in Malaysia (Neuroptera, Chrysopidae): the confluence of citizen scientist, online image database and cybertaxonomy. Zookeys, 11, 1–11.

Wortley, A. H., Scotland, R. W. 2004. Synonymy, sampling and seed plant numbers. Taxon, 53, 478–480.

Yesson, C., Brewer, P. W., Sutton, T., Caithness, N., Pahwa, J. S., Burgess, M., Gray, W. A., White, R. J., Jones, A. C., Bisby, F. a, Culham, A. 2007. How global is the global biodiversity information facility? PLoS One, 2, e1124.

Zauner, H. 2009. Evolving e-taxonomy. BMC Evol. Biol., 9, 141.

III. Correct Names for some of the Closest Relatives of *Carica papaya* L.: A Review of the Mexican/Guatemalan Genera *Jarilla* and *Horovitzia* [§]

Fernanda Antunes Carvalho[*] and Susanne S. Renner

Systematic Botany and Mycology, Ludwig-Maximilians-Universität München, Menzinger Strasse 67, D-80638 Munich, Germany

*Corresponding author: Fernanda Antunes Carvalho (antunesfc@gmail.com)

[§] published in: *PhytoKeys* 29: 63–74. doi:10.3897/phytokeys.29.6103

Keywords: Caricaceae • nomenclature • typification • epitypification • papaya sister clade

Abstract

Using molecular data, we recently showed that *Carica papaya* L. is sister to a Mexican/Guatemalan clade of two genera, *Jarilla* Rusby with three species and *Horovitzia* V.M. Badillo with one. These species are herbs or thin-stemmed trees and may be of interest for future genomics-enabled papaya breeding. Here we clarify the correct names of *J. heterophylla* (Cerv. ex La Llave) Rusby and *J. caudata* (Brandegee) Standl., which were confused in a recent systematic treatment of *Jarilla* (Mcvaugh 2001). We designate epitypes for both, provide weblinks to type specimens, a key to the species of *Jarilla* and *Horovitzia*, and notes on their habitats and distribution.

Introduction

The family Caricaceae Dumort. comprises 34 species and one formally named hybrid in currently six genera. A molecular phylogeny that included all species revealed that *Carica papaya* L. (the only species in the genus *Carica*) is sister to a clade of four species endemic to Mexico and Guatemala (Carvalho and Renner 2012). The discovery that the closest relatives of *C. papaya* are three herbs in the genus *Jarilla* Rusby and a thin stemmed tree, *Horovitzia cnidoscoloides* (Lorence & R. Torres) V. M. Badillo, has implications for plant breeders, who have so far tried in vain to cross papaya with tree species in the genus *Vasconcellea* A. St.-Hil., known as the highlands papayas. To facilitate communication among researchers from different fields, and since full-genome sequencing of the species of *Jarilla* and *Horovitzia* is ongoing (R. Ming, Urbana-Champaign, personal communication, Aug. 2013), we here provide a conspectus of the four species that are the closest relatives of papaya and clean up a nomenclatural confusion involving two names in the genus *Jarilla*.

We start with the nomenclatural issues, then provide a key to the four species, and end with brief comments on the range and habitat of each species.

Nomenclature of *Jarilla*

Pablo de La Llave (1832), a director of the National Museum of Natural History of Mexico, was the first to describe one of the unusual herbaceous Caricaceae that are today placed in *Jarilla*. He had access to fruiting specimens only and based his description of the flowers on notes made by Vicent Cervantes, a professor of botany in Mexico City and one of the founders of that city's botanical garden in 1788. La Lave gave his new species the epithet *"heterophilla"* [sic!] to refer to its variably shaped leaves. To mark the distinctness of the new species, he placed it in a separate genus, *Mocinna*, honoring the Mexican naturalist José Mariano Mociño. Unfortunately, this overlooked that Lagasca in 1816 had already described an Asteraceae genus by that name. Soon thereafter, George Bentham (1839) described the same species as *Carica nana*, based on an unnumbered Hartweg specimen (Fig. 1) collected in 1836 in Léon (Guanajuato, Mexico). The holotype at K (Fig. 1) bears the number *288* on its label, a number corresponding to the page of *Plantae Hartwegianae* on which *C. nana* was described. Diaz-Luna and Lomeli-Sención (1992), in their revision of *Jarilla*, cite this collection as *Hartweg 255* (K), probably due to a misreading of 288 for 255.

The second herbaceous Caricaceae species was named in March 1894 by Townshend S. Brandegee, who described *Carica caudata* from the Cape region of Baja California, Mexico, based on a plant he collected the year before (Fig. 2). In August of the same year, José Ramírez, unaware of Brandegee's publication, described a new variety of the first herbaceous Caricaceae, *M. heterophylla* La Llave, naming it varietas *sesseana*, based on living plants from Guanajuato and Jalisco. Unfortunately, he appears to have made no herbarium specimens, but only two beautiful plates showing the typical variety and var. *sesseana* (Fig. 3). Comparison of the plate of var. *sesseana* and the holotype of *C. caudata* leaves no doubt that these names refer to the same species, and we therefore agree with previous assessments (Diaz-Luna and Lomeli-Sención 1992, Badillo 1993) that they are synonyms.

Realizing that *Mocinna* La Llave was a younger homonym of *Mocinna* Lag., Henry Hurd Rusby (1921) proposed the substitute name *Jarilla*, derived from the Spanish vernacular name Jarrila, for *M. heterophylla*. He also up-ranked var. *sesseana* as a separate species, *Jarilla sesseana* (Ramírez) Rusby. We agree with Diaz-Luna and Lomeli-Sención (1992) and McVaugh (2001) that Rusby's publication of the substitute name *Jarilla* meets the requirement for valid publication and that Ivan M. Johnston's (1924) slightly later publication of the name *Jarrilla* (the correct Spanish spelling) to replace *Mocinna* is a superfluous name. At around the same time, Standley (1924) realized that *Carica caudata* Brandegee belonged in *Jarilla* and was in fact an older name for *J. heterophylla* var. *sesseana* Ramírez (= *Jarilla sesseana* (Ramírez) Rusby), and accordingly changed the name to *J. caudata*. He also described a third herbaceous species of Caricaceae, *Jarilla chocola* Standley, based on two collections made in 1935 from Sonora, Mexico (Standley 1937).

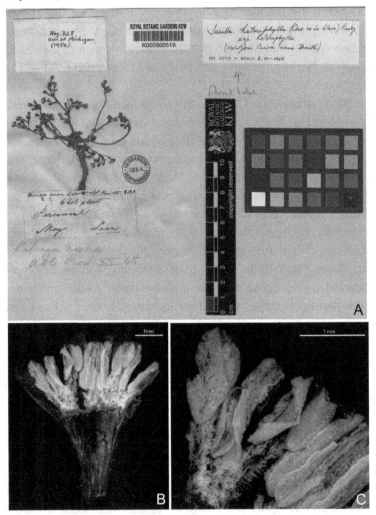

Fig. 1. Holotype of *Carica nana* Benth. **A** Specimen in K (http://www.kew.org/herbcatimg/202388.jpg); **B** Photo of an opened flower showing the arrangement of the anthers and the pistillode (*arrow*); **C** Close-up of the anthers. Filaments are densely covered by moniliform trichomes. B and C were taken by the first author in K.

Thus, by 1937 it was clear there were three species of *Jarilla* and also what their correct names were. In their revision of the genus, Diaz-Luna and Lomeli-Sención (1992) designated plate II of Ramírez (1894; our Fig. 3 left-hand plate) as the lectotype of *J. heterophylla* var. *sesseana* and plate V as the neotype of var. *heterophylla* (our Fig. 3 right-hand plate). Unfortunately, the most recent study of *Jarilla*, that of Rogers McVaugh (2001), synonymized the two taxa distinguished by Ramírez and used Bentham's *Carica nana* (= *Jarilla nana* (Benth.) McVaugh) as the oldest name for the second species of *Jarilla* (the one described by Brandegee as *C. caudata*). Such treatment is surprising given the different leaves and fruits of Ramírez's two varieties (our Fig. 3), and indeed McVaugh, might not be completely sure about that, because he writes (2001: 469), "In the following I have drawn heavily upon the work of Diaz-Luna and Lomelí-Sención, whose personal observations of these interesting species greatly increased our knowledge of them, and have indeed provided almost all the available information about the living plants. Errors introduced here, as a result of faulty translation or interpretation of the work of these authors, or otherwise, are solely my responsibility."

We agree with Diaz-Luna and Lomeli-Sención (1992) and the earlier workers cited above that *Jarilla heterophylla* var. *heterophylla* is the oldest name for Bentham's *Carica nana*, while var. *sesseana* is a younger synonym of *Carica caudata*. We have accordingly up-dated the names of our previous *Jarilla heterophylla* and *J. nana* sequences in GenBank (Carvalho and Renner 2012; all of which are vouchered). Together, the descriptions of Ramírez (1894), Brandegee (1894), Rusby (1921), Johnson (1924), Standley (1924), and Diaz-Luna and Lomeli-Sención (1992) provide a clear idea of the morphological distinctions of the two species: *Jarilla caudata* has rounded to ovate or deltoid (never hastate) leaves, c. 1 cm (rarely longer) male flowers, and 10 cm long fruits that are narrowed at the base with five horn-like appendages, each 3–6 cm long (Fig. 4). *Jarilla heterophylla* has hastate leaves, 0.5 cm long male flowers, and c. 3 cm long fruits with short and thick appendages as shown in Fig. 5.

Fig. 2. Holotype of *Carica caudata* Brandegee (http://ucjeps.berkeley.edu/new_images/UC108333.jpg)

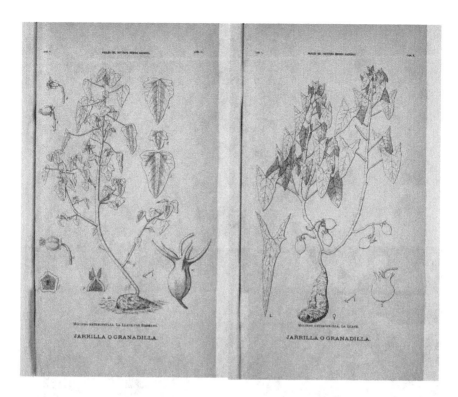

Fig. 3. The two varieties of *Mocinna heterophylla* La Llave. **Left plate** shows the lectotype of *Mocinna heterophylla* var. *sesseana* Ramírez. **Right plate** shows the neotype of *Mocinna heterophylla* var. *heterophylla*, both designated by Díaz and Lomeli-Sención (1992). Plates reproduced from Ramírez (1894).

Fig. 4. *Jarilla caudata* (Brandegee) Standl. **A** Epitype of *Mocinna heterophylla* La Llave (*F.A. Carvalho 2240*, M). **B** Habit. **C** Male inflorescence. **D** Staminate flower. **E** Pistillate flower. **F** Fruits. **G** Ovary unilocular and seeds **H** Tuber.

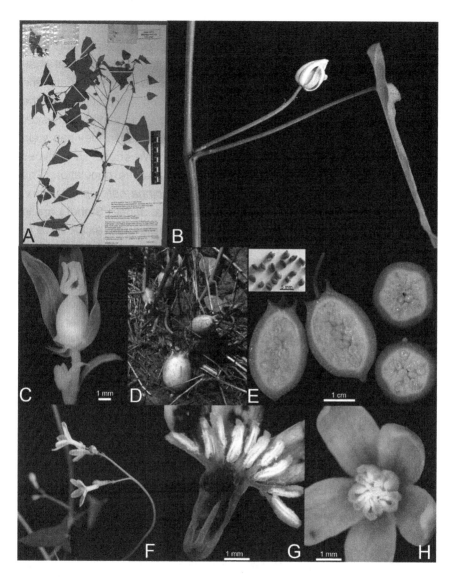

Fig. 5. *Jarilla heterophylla* (Cerv. ex La Llave) Rusby. **A** Epitype of *Mocinna heterophylla* La Llave var. *sesseana* Ramírez (*F.A. Carvalho 2239*, M). **B** Female inflorescence (uniflora). **C** Female flower showing the short appendages at the base of the ovary. **D–E** Fruits and seeds. **F** Male inflorescence. **G–H** Staminate flowers.

41

To fix the usage of the two names more reliably, we designate below epitypes to serve as interpretative specimens for plates II and V of Ramírez (1894), following Art. 9.8 of the Melbourne Code (McNeill et al. 2012). The plates published by Ramírez fail to include staminate and pistillate flowers for both species and therefore do not precisely fix the application of the names of these dioecious species. In addition, physical specimens also can help in evolutionary studies because they can yield DNA that may be used in future comparisons. We chose as epitypes complete male and female specimens from the same population. The epitypes are deposited in M. Isoepitypes of *Mocinna heterophylla* Cerv. ex La Llave var. *sesseana* (=*Jarilla caudata* (Brandegee) Standl.) are in MEXU and NY. Isoepitypes of *Mocinna heterophylla* Cerv. ex La Llave (= *Jarilla heterophylla* (Cerv. ex La Llave) Rusby) are in MEXU, NY and K.

The four species in the *Jarilla*/*Horovitzia* clade can be distinguished from all other Caricaceae and from each other, using a combination of the plastid markers *trnL-trnF* and *psbA-trnH* (Carvalho & Renner, 2012; GenBank accessions JX091966, JX091977, JX091975, JX091978, JX092054, JX092064, JX092065, JX092066).

Key to the species of *Jarilla* and *Horovitzia*

1a. Small tree, completely covered by stinging hairs   *Horovitzia cnidoscoloides*
1b. Herb, glabrous or pubescent, but never with stinging hairs. 2
2a. Erect herb. Leaves lobate, rarely entire. Ovary and mature fruits with 5 longitudinal wings. Female flowers 7–9 mm long. Male flowers 5–9 mm long . . . .
. . . .*Jarilla chocola*
2b. Procumbent herb, sometimes using understory plants for support. Leaves entire, rarely lobed. Ovary and young fruits with 5 basal appendages, but not winged. Female flowers 5-15 mm long. Male flowers 4-12 mm long. 3
3a. Mature fruits 6–30 cm long with 5 horn-like basal appendages 3–6 cm long. Seeds black, 4–5.5 mm long. Male flowers in general >1 cm (1–1.7 cm) . . . .
. . . .*Jarilla caudata*
3b. Mature fruits 2–4 cm long with 5 curved basal appendages 0.5–2 cm long. Seeds light brown, 2.5–3.5 mm long. Male flowers in general <1 cm (0.3–0.8 cm) . . . .
. . . .*Jarila heterophylla*

Epitypification and comments on morphology and habitats

Horovitzia cnidoscoloides (Lorence & R. Torres) V.M. Badillo, Rev. Fac. Agron. (Maracay) 43: 104, 1993.
Carica cnidoscoloides Lorence & R. Torres, Syst. Bot. 13(1): 107–109, f.1. 1988.
　　Type: Mexico. Oaxaca: Ixtlan, Sierra de Juárez, 9 March 1986, *R. Torres & P. Teonorio 8168* (holotype: MEXU, a photo in GUADA; isotypes: BM, MO [MO-193213], NY[00112155]). Mexico. Oaxaca. Type locality, 25 May 1883, *T. Cedillo & Lorence 2347* (paratype: MEXU, a photo in GUADA, a duplicate in MO); 4 Ago 1985, *Lorence et al. 4733* (paratype: MEXU, a duplicate in BM); 9 Mar 1985, *C. Torres & L. Tenorio 8167* (paratype: MEXU); 27 Ago 1986, *C. Torres & L. Tenorio 8760* (paratype: MEXU).

Horovitzia cnidoscoloides is a small tree, 0.5–4 m tall endemic to Sierra de Juarez in Oaxaca, Mexico. It occurs in cloud forests from 800 to 1600 m above sea level. Unusual features are subcapitate stigma, and stinging hairs covering the entire plant.

Jarilla chocola Standl. Publ. Field Mus. Nat. Hist., Bot. Ser. 17: 200, 1937.
　　Type: Mexico. Sonora: Chihuahua, Guasarema, Rio Mayo, 10 August 1936, *H. S. Gentry 2366* (holotype: F; isotypes: GUADA photo, K [K000500520], S [S-G-3434]). Mexico. Sonora: San Bernardo, Rio Mayo, 14 August 1935, *H. S. Gentry 1624* (paratype: F, a photo in GUADA, a duplicate in MEXU).

Jarilla chocola is an erect herb, with mostly lobate leaves and fruits with 5 longitudinal wings. The species occurs at low altitudes (100–1300 m) along the Pacific Coast from Sonora to El Salvador.

Jarilla caudata (Brandegee) Standl., Contr. U.S. Natl. Herb. 23(4): 853, 1924 (Fig. 4).

Carica caudata Brandegee, Zoe 4: 401. 1894.

Type: Mexico. Baja California Sur: Corral de Piedra, September 1893, *Brandegee s.n.* (holotype: UC[UC108333]).

Mocinna heterophylla var. *sesseana* Ramírez, Anales Inst. Med.-Nac. Mexico 1: 207, 1894.

Type: Plate II of Ramírez, 1894 (lectotype designated by Diaz-Luna & Lomeli-Sención 1992: 81). Mexico, Jalisco, Zacoalco de Torres, Las Moras, 5 June 2013, *F. A. Carvalho 2239* (epitype, designated here: M; isoepitypes: MEXU, NY).

Jarilla sesseana (Ramírez) Rusby, Torreya 21: 47, 1921.

Jarilla caudata is morphologically and phylogenetically closely related to *J. heterophylla*. Their main distinguishing features are the fruits, which in *J. caudata* can attain a length of 30 cm, having a smooth surface and 5 long, horn-like appendages (3–6 cm long). Other differences are given in the key. The species occurs in deciduous forests and fields of Baja California and central Mexico from 1500 to 1800 m above sea level.

Jarilla heterophylla (Cerv. ex La Llave) Rusby, Torreya 21(3): 50, 1921 (Fig. 5).

Mocinna heterophylla Cerv. ex La Llave, Reg. Trim. 1(3): 351, 1832.

Type: Plate V of Ramírez, 1894 (neotype, designated by Diaz-Luna & Lomeli-Sención 1992: 88). Mexico, Jalisco, Zacoalco de Torres, Las Moras, 5 June 2013, *F. A. Carvalho 2240* (epitype, designated here: M; isoepitypes: MEXU, NY, K).

Carica nana Benth., Pl. Hartw. 288. 1849.

Type: Mexico. Guanajuato, Leon, *K. T. Hartweg s.n.* (holotype K [K000500519]; isotype: G-DC n.v.).

Papaya nana (Benth.) A. DC., Prodr. 15(1): 415, 1864.

Jarilla nana (Benth.) McVaugh, Fl. Novo-Galiciana 3: 475, 2001.

For differences from *Jarilla caudata* see under that species and in the key. *Jarilla heterophylla* occurs in oak forests, deciduous forests, and abandoned fields of central Mexico at 1500 to 2700 m above sea level.

Weblinks to type specimens

Carica caudata Brandegee, holotype:
http://ucjeps.berkeley.edu/new_images/UC108333.jpg [accessed 30.07.2013]

Carica cnidoscoloides Lorence & R. Torres, isotypes:
http://www.tropicos.org/Image/11116 [accessed 11.08.2013]
http://sweetgum.nybg.org/vh/specimen.php?irn=707429 [accessed 11.08.2013]

Carica nana Benth., holotype:
http://www.kew.org/herbcatimg/202388.jpg [accessed 30.07.2013]

Jarilla chocola Standl., isotypes:
http://apps.kew.org/herbcat/getImage.do?imageBarcode=K000500520 [accessed 11.08.2013]
http://andor.nrm.se/kryptos/fbo/kryptobase/large/S-G-003001/S-G-3434.jpg [accessed 11.08.2013]

Mocinna heterophylla Cerv. ex La Llave, epitype:
http://herbaria.plants.ox.ac.uk/bol/caricaceae [accessed 11.10.2013]

Mocinna heterophylla var. *sesseana* Ramírez, epitype:
http://herbaria.plants.ox.ac.uk/bol/caricaceae [accessed 11.10.2013]

Acknowledgements
We thank J. F. Barêa Pastore for discussion of the nomenclature of *Jarilla*, and the herbaria K, MO, NY, S, and UC for providing open access to type images. The first author is supported by a fellowship from Conselho Nacional de Desenvolvimento Científico e Tecnológico (CNPq 290009/2009-0), and additional funding came from the German Research Foundation (DFG RE 603/13).

References

Badillo VM (1993) Caricaceae, segundo esquema. Revista de la Faculdad de Agronomia de la Universidad Central de Venezuela 43:1–111.

Bentham G (1839) Plantae Hartwegianae. London, Facsimile edition: Lehre J. Cramer 1970. http://www.biodiversitylibrary.org/page/796739#page/288/mode/1up [accessed 30.07.2013]

Brandegee TS (1894) Additions to the flora of the Cape region of Baja California. II. Zoe 4(4): 398–407. http://www.biodiversitylibrary.org/page/568039#page/97/mode/1up [accessed 30.07.2013]

Carvalho FA, Renner SS (2012) A dated phylogeny of the papaya family (Caricaceae) reveals the crop's closest relatives and the family's biogeographic history. Molecular Phylogenetics and Evolution 65(1): 46–53. doi: 10.1016/j.ympev.2012.05.019

Diaz-Luna CL, Lomelí-Sención JA (1992) Revisión del género *Jarilla* Rusby (Caricaceae). Acta Botánica Mexicana 20: 77–99. http://www.redalyc.org/pdf/574/57402010.pdf [accessed 30.07.2013]

Johnston IM (1924) Taxonomic notes concerning the American Spermatophytes. New or otherwise noteworthy plants. Contributions from the Gray Herbarium of Harvard University 70: 69–87. http://www.biodiversitylibrary.org/page/39944720#page/77/mode/1up [accessed 30.07.2013]

Kyndt T, Van Droogenbroeck B, Romeijn-Peeters E, Romero-Motochi JP, Scheldeman X, Goetghebeur P, Van Damme P, Gheysen G (2005) Molecular phylogeny and evolution of Caricaceae based on rDNA internal transcribed spacers and chloroplast sequence data. Molecular Phylogenetics and Evolution 37(2): 442–59. doi: 10.1016/j.ympev.2005.06.017

La Llave P (1832) Descripcion de alguns géneros y especies nuevas de vegetales. Registro Trimestre ó Coleccion de Memorias de Historia, Literatura, Ciencias y Artes1(3): 345–358. http://www.biodiversitylibrary.org/page/14631797#page/373/mode/1up [accessed 30.07.2013]

Lorence DH, Colín RT (1988) *Carica cnidoscoloides* (sp. nov.) and sect. *Holostigma* (sect. nov) of Caricaceae from Southern Mexico. Systematic Botany 13(1): 107–110.

McNeill J, Barrie FR, Buck WR, Demoulin V, Greuter W, Hawksworth DL, Herendeen PS, Knapp S, Marhold K, Prado J, Prud'homme van Reine WF, Smith JF, Wiersema JH, Turland NJ (2012) International Code of Nomenclature for Algae, Fungi, and Plants (Melbourne Code): adopted by the Eighteenth International Botanical Congress Melbourne, Australia, July 2011. Regnum Vegetabile 154. Koeltz Scientific Books. http://www.iapt-taxon.org/nomen/main.php [accessed 30.07.2013]

McVaugh R (2001) Caricaceae. In: Anderson WR (Ed) Flora Novo-Galiciana. A descriptive account of the vascular plants of Western Mexico. Vol. 3 Ochnaceae to Losaceae. The University of Michigan Press, Ann Arbor, 461–477.

Ramírez J (1894) La *Mocinna heterophylla*. Nuevo género de las papayáceas. Anales del Instituto Médico Nacional 1(5): 205–212, pl. II–V. http://archive.org/stream/analesdelinstitu01inst#page/206/mode/2up [accessed 30.07.2013]

Rusby HH (1921) A strange fruit. Torreya 21(3): 47–50. http://www.biodiversitylibrary.org/item/100133#page/243/mode/1up [accessed 30.07.2013]

Standley PC (1924) Caricaceae. In: Standley PC (Ed) Tree and shrubs of Mexico (Passifloraceae-Scrophulariaceae). Contributions from the United States National Herbarium 23(4): 849–853. http://www.biodiversitylibrary.org/page/375754#page/939/mode/1up [accessed 30.07.2013]

Standley PC (1937) Caricaceae. In: Standley PC (Ed) Studies of American Plants–VII. Publications of the Field Museum of Natural History, Botanical Series 17(2): 200–202.

IV. A Dated Phylogeny of the Papaya Family (Caricaceae) Reveals the Crop's Closest Relatives and the Family's Biogeographic History [§]

Fernanda Antunes Carvalho, Susanne S. Renner

Systematic Botany and Mycology, University of Munich (LMU), Menzinger Strasse 67, 80638 Munich, Germany

* Corresponding author: antunesfc@gmail.com (F.A. Carvalho)

[§] published in: *Molecular Phylogenetics and Evolution* 65(1), 46–53, October 2012
DOI: doi:10.1016/j.ympev.2012.05.019,

Keywords: Long distance dispersal • molecular clocks • historical biogeography • Panamanian Isthmus

Abstract

Papaya (*Carica papaya*) is a crop of great economic importance, and the species was among the first plants to have its genome sequenced. However, there has never been a complete species-level phylogeny for the Caricaceae, and the crop's closest relatives are therefore unknown. We investigated the evolution of the Caricaceae based on sequences from all species and genera, the monospecific *Carica*, African *Cylicomorpha* with two species, South American *Jacaratia* and *Vasconcellea* with together c. 28 species, and Mexican/Guatemalan *Jarilla* and *Horovitzia* with four species. Most Caricaceae are trees or shrubs; the species of *Jarilla*, however, are herbaceous. We generated a matrix of 4177 nuclear and plastid DNA characters and used maximum likelihood (ML) and Bayesian analysis to infer species relationships, rooting trees on the Moringaceae. Divergence times were estimated under relaxed and strict molecular clocks, using different subsets of the data. Ancestral area reconstruction relied on a ML approach. The deepest split in the Caricaceae occurred during the Late Eocene, when the ancestor of the Neotropical clade arrived from Africa. In South America, major diversification events coincide with the Miocene northern Andean uplift and the initial phase of the tectonic collision between South America and Panama resulting in the Panamanian land bridge. *Carica papaya* is sister to *Jarilla/Horovitzia*, and all three diverged from South American Caricaceae in the Oligocene, 27 (22–33) Ma ago, coincident with the early stages of the formation of the Panamanian Isthmus. The discovery that *C. papaya* is closest to a clade of herbaceous or thin-stemmed species has implications for plant breeders who have so far tried to cross papaya only with woody highland papayas (*Vasconcellea*).

1. Introduction

Annual world production of *Carica papaya* is now >10 million tons, making papaya an extremely important fruit crop (Scheldeman *et al.*, 2011; FAOStat, 2011). Papaya is a vital source of vitamins for people in the humid tropics, and proteinases obtained from the milky latex extracted from unripe fruits are widely used in the food and

50

pharmaceutical industry (Krishnaiah *et al.*, 2002; Scheldeman *et al.*, 2007). The so-called highland papayas from the Andes (*Vasconcellea* spp.) regarded as underexploited crops, have a number of desirable traits, including disease resistance, cold tolerance, and latex with high proteolytic activity (Scheldemann *et al.*, 2011). Breeders have sought to introduce these features to *C. papaya* using traditional plant breeding techniques. However, although some experimental hybrids exhibit useful disease resistance, serious fertility barriers limit breeding efforts (Sawant, 1958; Dinesh *et al.*, 2007; Siar *et al.*, 2011). The commercial interest in papaya is reflected in more than 1300+ papers published on this species between 1970 and 2011 (*Web of Science*, accessed 25 September 2011), ca. half of them on papaya ring spot virus. Papaya was also among the first plants to have its genome sequenced, with the data revealing that it has relatively few lignin-synthetic genes (Ming *et al.*, 2008), fitting with its reduced woodiness. To take full advantage of this genome, for example to understand the evolution of lignin-related genes, requires knowing the closest relatives of papaya. Wild papaya populations are strictly dioecious, but cultivated and feral populations include individuals with a modified (mal-functioning) Y chromosome that bear bisexual flowers and are favored in plantations because of the particular fruit shape. An incipient Y chromosome has been shown to control sex in papaya (Liu *et al.*, 2004; Ming *et al.*, 2011; Wu *et al.*, 2010), and knowing the closest relatives of this crop will also help understand the origin of sex chromosomes in papaya and other Caricaceae.

The papaya family (Caricaceae) has an amphi-Atlantic distribution with two species in tropical Africa and 33 in Central and South America (Fig. 1). The family is currently divided into six genera of which *Carica* is one, with the only species *C. papaya* (Badillo, 1993, 2000). Both African species are large trees, one (*Cylicomorpha solmsii*) in West Africa, the other (*C. parviflora*) in East Africa (Fig. 1E and F). The monotypic genus *Horovitzia* (*H. cnidoscoloides*), endemic to Mexico, is a small tree with spongy thin stems covered with stinging hairs (Lorence and Colin, 1988; Badillo, 1993). The likewise Mexican (to Guatemalan) genus *Jarilla*

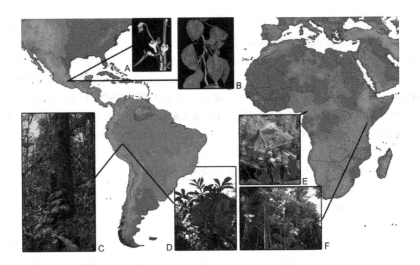

Fig. 1. Distribution of the Caricaceae, except for the cultivated *Carica papaya*. The insets show the Mexican/Guatemalan herbs A) *Jarilla nana* (photo: M. Olson) and B) *Jarilla heterophylla* (photo: F. A. Carvalho); C) the spiny trunk of the tree *Jacaratia digitata* from Cusco, Peru (photo: E. Honorio); D) habit of *Vasconcellea* cf. *microcarpa*, from eastern Andes, Cusco, Peru (alt.: 600 m; photo: E. Honorio); E) fruits of the African *Cylicomorpha solmsii* from Yaoundé, Cameroon (alt.: 900 m; photo: J. P. Gogue) and; F) the gregarious habitat of *Cylicomorpha parviflora* from Kiangombe hill, Kenya (alt.: 1600 m; photo: M. Nicholson).

(Fig. 1A and B) comprises three species of perennial herbs (McVaugh, 2001). *Jacaratia* (Fig. 1C) currently has eight tree species (including a suspected new species) occurring from southern Brazil to Mexico, and finally, *Vasconcellea* consists of 20 species, 19 of them trees (Fig. 1D) or shrubs, and one a climber. The highest species density of *Vasconcellea* is in the northwestern Andes. The sister group of Caricaceae is the Moringaceae, a family of 13 species of trees and shrubs from dry habitats in the Horn of Africa (seven species), Madagascar (two species), southwestern Africa (one species), and tropical Asia (three species; Olson, 2002). Large-scale dating efforts for angiosperms and Brassicales have placed the split between Caricaceae and Moringaceae in the Early Paleocene, at c. 65 Ma (Wikström *et al.*, 2001; Beilstein *et al.*, 2010; Bell *et al.*, 2010).

Several molecular-phylogenetic analyses of Caricaceae have been undertaken (Jobin-Decor *et al.*, 1997: using isozymes; Aradhya *et al.*, 1999: Rflps; Olson, 2002: nuclear ITS and plastid sequences; Van Droogenbroeck *et al.*, 2002: AFLPs, 2004: PCR-AFLPs; Kyndt *et al.*, 2005a: nuclear ITS and plastid sequences, Kyndt *et al.*, 2005b: AFLPs, PCR-RFLPs) but none included representatives of all genera. The evolutionary relationships within the family have therefore remained unclear. Wild *C. papaya* populations (referred to *Carica peltata* Hook. & Arn. until Badillo (1971) synonymized the name under *C. papaya*) are found from southern Mexico through Nicaragua and Belize to Guatemala and Costa Rica (Storey, 1976; Moreno, 1980; Manshardt, 1992; Manshardt and Zee, 1994; Morshidi *et al.*, 1995; Paz and Vázquez-Yanes, 1998; Coppens d'Eeckenbrugge *et al.*, 2007; Brown *et al.*, 2012), fitting with a Mexican or Central American domestication of papaya as advocated by De Candolle (1883), Solms-Laubach (1889), and Vavilov (1940 [1992]). Prance (1984) suggested domestication on the southwestern side of the Andes, while Brücher (1987) briefly favored an Amazonian origin (but see Brücher, 1989; Prance and Nesbitt, 2005). So far, there is no direct archaeological evidence because papaya cannot be identified from phytoliths, and pollen grains are rarely found (D. Piperno, Smithsonian Tropical Research Institute, Panama; email to S. Renner, 17 October 2010). An isozyme analysis of numerous papaya accessions, while revealing limited genetic diversity, showed wild papaya plants from Yucatán, Belize, Guatemala, and Honduras more related to each other than to domesticated plants from the same region (Morshidi *et al.*, 1995).

To understand the biogeography of Caricaceae and to identify the sister clade of papaya, we compiled nuclear and plastid DNA sequences from all of the family's extant species and then used a molecular clock approach to infer the divergence times of major groups. It is well established that the land bridge dividing the Pacific from the Atlantic Ocean was fully established by 3–3.5 Ma (O'Dea *et al.*, 2007; Farris *et al.*, 2011). However, previous inferences about the duration of the gradual shoaling process were based mainly on the divergence times of marine organisms and are relatively imprecise. New evidence on when isthmus formation began comes from geological sequences in Panamanian and Colombian sedimentary basins that are 14.8–12.8 Ma old (Farris *et al.*, 2011). The latter authors suggested that collision between Central America and South America initiated at 23–25 Ma, when South America first impinged upon Panamanian arc crust.

To summarize, the specific goals of this study were to (i) infer the closest relatives of papaya, (ii) produce a dated phylogeny for the Caricaceae, and (iii), to the extent possible, relate the diversification of Caricaceae to geological or climatic events both in Africa and in Central and South America.

2. Materials and Methods

2.1. Taxon sampling, DNA sequencing, alignment

We sequenced 36 accessions representing the 34 species recognized in the latest revision of Caricaceae (Badillo, 1993), as well as a new species (leg. E. Honorio 1365). *Vasconcellea* x *heilbornii*, a sterile hybrid between *V. cundinamarcensis*, *V. stipulata*, and *V. weberbaueri* (maintained in cultivation as a parthenogenetic clone), was excluded from this study (Aradhya *et al.*, 1999; Van Droogenbroeck *et al.*, 2002, 2004; Scheldemann *et al.*, 2011). Seven of the 13 species of Moringaceae were chosen as outgroups for rooting purposes (Olson, 2002). Appendix S1 lists DNA sources with voucher information, species names with authorities, GenBank accession numbers, and the general distribution of each species.

Total genomic DNA was extracted from 1–23 mg of leaf tissue from herbarium specimens or, more rarely, silica-dried leaves, using a commercial plant DNA extraction kit (NucleoSpin, Macherey-Nagel, Düren, Germany) according to manufacturer protocols. Polymerase chain reactions (PCR) followed standard protocols, using Taq DNA polymerase and 22 different primers (Appendix S2). PCR products were purified with the ExoSap clean-up kit (Fermentas, St. Leon-Rot, Germany), and sequencing relied on Big Dye Terminator kits (Applied Biosystems, Foster City, CA, USA) and an ABI 3130 automated sequencer. In all, 228 chloroplast sequences (*trnL–trnF*, *rpl20–rps*12, *psbA–trnH* intergenic spacers, *matK* and *rbcL* genes) and 35 nuclear sequences (from the Ribosomal DNA internal transcribed spacers ITS1 and ITS 2, plus the intervening 5.8 S gene) were newly generated for this study (Appendix S1). All new sequences were BLAST-searched in GenBank and then first aligned using MAFFT vs. 6 (http://mafft.cbrc.jp/alignment/server/) using defaults parameters.

The Q-INS-i multiple alignment strategy was chosen for the ITS sequences because it considers secondary structure and is recommended for alignments of highly diverged ncRNAs (Katoh and Toh, 2008). Minor alignment errors were manually adjusted in MacClade vs. 4.06 (Maddison and Maddison, 2005). In order to remove poorly aligned positions, alignments were exported to a server running Gblocks vs. 0.91b (http://molevol.cmima.csic.es/castresana/Gblocks_server.html) with the less stringent options selected (Castresana, 2000).

2.2. Phylogenetic analyses and molecular clock dating

In the absence of statistically supported (i.e., >70% bootstrap support) topological contradictions from maximum likelihood (ML) tree searches (below), the chloroplast and nuclear data were combined, yielding a matrix of 4711 aligned characters. Phylogenetic trees were estimated using ML and Bayesian optimization, the former in RAxML (Stamatakis, 2006), using the RAxML GUI vs. 0.93 (Silvestro and Michalak, 2011), the latter in BEAST vs. 1.6.1 (Drummond and Rambaut, 2007). The ML analyses used the GTR + Γ model with six rate categories, with independent models for each data partition and model parameters estimated over the duration of specified runs. Statistical support came from bootstrapping under the same model, with 100 replicates. BEAST analyses relied on the uncorrelated lognormal relaxed clock, the GTR + Γ substitution model with four rate categories, and a Yule tree prior. Monte Carlo Markov chains (MCMC) were run for 10 million generations, with parameters sampled every 1000 generations. Log files were then analyzed with Tracer vs. 1.5 (http://beast.bio.ed.ac.uk/) to assess convergence and to confirm that the effective sample sizes for all parameters were larger than 200, indicating that MCMC chains were run long enough to reach stationarity. After discarding 10% of the saved trees as burn-in, a maximum credibility tree based on the remaining trees was produced using TreeAnnotator (part of the BEAST package) and FigTree vs. 1.3.1 (http://tree.bio.ed.ac.uk).

There are no Caricaceae or Moringaceae fossils, and we therefore resorted to secondary calibration of our clock models. The age of the Caricaceae/Moringaceae node has been estimated in three large-scale studies that used fossil calibrations (Wikström *et al.*, 2001: 59 Ma (58–61); Beilstein *et al.*, 2010: 69 Ma (105–38); Bell *et al.*, 2010: 67 Ma (45–86)). We assigned this node a normally distributed prior with a mean of 65 Ma and a standard deviation of 2 Ma, reflecting the age range estimated by the three studies. Two alignments of 26 plant accessions were used for dating, one consisting of the plastid genes *mat*K and *rbc*L ('slow data'), which exhibit relatively low nucleotide substitution rates; the other consisting of all six markers ('fast data'). Dating in BEAST used the same substitution model and tree prior as used in the phylogenetic analysis and either an uncorrelated relaxed clock or a strict clock model (the latter only for the 'slow data' matrix).

2.3. Biogeographic analyses

Species ranges were coded using the monograph of Badillo (1971) and more recent studies (Diaz-Luna and Lomeli-Sención, 1992; Badillo, 1993; McVaugh, 2001; Lleras, 2010). The species ranges were assigned to one of the three regions: (1) Africa; (2) South America including southernmost Panama, and (3) Central America from Costa Rica to Mexico (Fig. 2). Two widespread species (*Vasconcellea cauliflora* and *Jacaratia spinosa*) were coded as 'South America/Central America'. Because wild populations of papaya occur from Mexico south through Belize and the Petén region of Guatemala, it was coded as 'Central America'. In an alternative run, we coded papaya as 'unknown' (as done in other studies of cultivated crops; Sebastian *et al.*, 2010) to reflect its anthropogenic distribution range. Moringaceae were coded as 'Africa'.

Ancestral area reconstruction (AAR) relied on the dispersal-extinction-cladogenesis (DEC) ML approach implemented in LAGRANGE (Ree *et al.*, 2005; Ree and Smith, 2008). Python input scripts were generated using an online tool (http://www.reelab.net/lagrange/configurator/index), with the maximum number of ancestral areas constrained to two. No dispersal constraints were defined. DEC analyses were run either with the ML tree (above), with the root age set to 65 Ma, or with an ultrametric tree obtained under a strict clock model and the 'slow data'.

3. Results

3.1. Evolutionary relationships in Caricaceae and the closest relatives of papaya

Phylogenies obtained with ML (Fig. 2) or Bayesian optimization (Fig. S1) are congruent and show that all four genera with more than one species are monophyletic. The deepest divergence in the family is between the clade containing the two African species (*Cylicomorpha solmsii* and *C. parviflora*) and the Neotropical clade of the remaining Caricaceae. In the latter, the first divergence involves *Jacaratia* and *Vasconcellea* on the one hand and the three Mexican/Guatemalan genera on the other, namely *Carica, Jarilla*, and *Horovitzia* (Fig. 2).

Within *Jacaratia*, *J. spinosa* (a widespread species throughout the Neotropical region) is sister to a clade comprising the remaining species (Fig. 2). The two Central American species, *J. dolichaula* and *J. mexicana*, are embedded in a South American clade, implying that they reached Central America from South America (Figs. 2 and S1). A suspected new species (*Jacaratia* Honorio 1365) from the eastern Andean foothills (alt. 1100–1300 m) in Peru is sister to *J. chocoensis*, an endemic of the Colombian Chocó region (alt. 600–800 m) to the west of the Andes. *Vasconcellea* comprises two species groups (with high statistical support, Figs. 2 and S1): a small clade consisting of four species from the western Andes (*V. candicans*), the coastal region of Central Chile (*V. chilensis*), and Peru, Bolivia, Brazil and southern South America (*V. glandulosa; V. quercifolia*). The larger clade comprises the remaining *Vasconcellea* species, which either have a narrow distribution in the western Andes in Peru and Ecuador (e.g., *V. stipulata, V. parviflora, V. weberbaueri*) or are more widely distributed, as for example *V. cauliflora*, which ranges from Mexico to Ecuador.

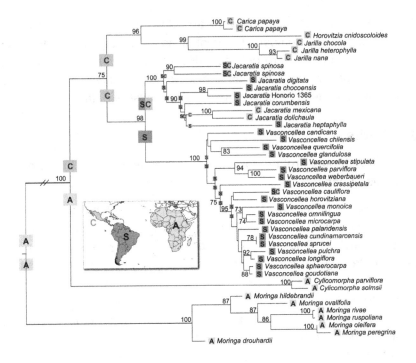

Fig. 2. Maximum likelihood tree for 37 accessions representing 34 species of Caricaceae based on 4711 aligned nucleotides of nuclear and plastid sequences. Bootstrap values at nodes (>70) are based on 100 replicates. Letters on branches represent the inferred areas for the Caricaceae obtained from the Dispersal-Extinction-Cladogenesis analysis, with *C. papaya* coded as 'Central America' (see text for alternative coding of this species). The area coding is shown in the inset and before each species name, with A meaning Africa, S South America, and M Mexico/Guatemala. (For interpretation of the references to color in this figure legend, the reader is referred to the web version of this article.)

3.2. Molecular clock dating

Divergence times inferred for biogeographically interesting events under strict clock and relaxed clock models are shown in Table 1. The ages inferred from the 'fast data' matrix lie within the confidence intervals of those inferred from the 'slow data'. Analysis of the log file generated under the relaxed clock model applied to the slow data gave ucld.mean and ucld.stdev parameters of 0.0003 and 0.492, respectively, suggesting that the *matK-rbcL* data are clock like

58

(Drummond *et al.*, 2007). Figure 3 shows a chronogram from these data. The divergence between the African *Cylicomorpha* clade and the Neotropical Caricaceae occurred 35.5 (28–43) Ma ago, during the Late Eocene. *Cylicomorpha solmsii* and *C. parviflora* shared a most common ancestor during the Pliocene-Pleistocene periods, around 2.8 (0.6–5.2) Ma. The split between the mostly South American *Vasconcellea/Jacaratia* clade and the Mexican/ Guatemalan papaya clade occurred in the Oligocene, ca. 27 (21.9–33) Ma ago, and *C. papaya* then diverged from its closest relatives an estimated 25 (19.5–31) Ma ago. The split between *Horovitzia cnidoscoloides* and the three *Jarilla* species is dated to the Early Miocene c. 18 (13.4–23.3) Ma and the *Jarilla* crown group to around 7 (4–10) Ma. The most recent common ancestor of the mostly southern South American *Jacaratia* and *Vasconcellea* clade dates to the Early Miocene, c. 19 (14.4–23.7) Ma ago, and the genera then started to diversify during the middle Miocene around 13 Ma (Fig. 3 and Table 1).

Table 1. Estimated node ages for selected divergence events under a strict clock model and a relaxed clock model using different datasets as indicated above columns. Ages are in million years, and the values in brackets are the 95% posterior probability intervals.

Nodes of interest	Molecular clock model		
	Slow data (*matK*, *rbcL*)		Fast data (ITS + plastid markers)
	Strict clock	Relaxed clock	Relaxed clock
1. Caricaceae	35.5 (28.1 – 43.1)	43.1	40.8 (29.2 – 52.6)
2. Neotropical clade	27.5 (21.9 – 33.4)	34.2 (19.6 – 50.4)	32.5 (23.5 – 42)
3. *Jacaratia/Vasconcellea*	18.8 (14.4 – 23.7)	23.8 (12.9 – 35.8)	22.4 (16.2 – 29.6)
4. *Jacaratia*	12.7 (8.6 – 17.1)	15.7 (7.6 – 25.1)	16.8 (11.3 – 22.5)
5. *Vasconcellea*	13.7 (9.9 – 17.3)	17 (9.1 – 25.9)	13.1 (9 – 17.5)
6. *Carica papaya*	25.1 (19.5 – 31.1)	30.4 (17.9 – 48.8)	25.3 (17.6 – 33)
7. *Horovitzia*	18.3 (13.4 – 23.3)	20.8 (10.5 – 32.9)	17.8 (11.8 – 24)
8. *Jarilla*	7.1 (4 – 10.4)	8.4 (3.3 – 14.8)	6 (3.5 – 8.8)
9. *Cylicomorpha*	2.8 (0.6 – 5.2)	3.7 (0.4 – 8.2)	3 (1.3 – 4.9)

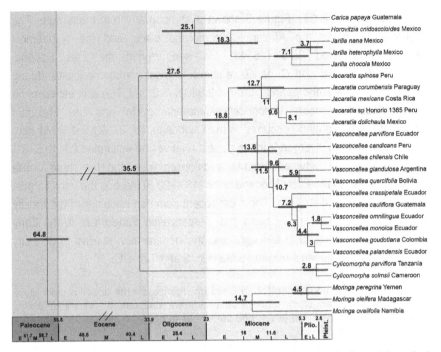

Fig. 3. A chronogram for the Caricaceae obtained under a strict clock model applied to two plastid genes *matK* and *rbcL*. The node bars indicate 95% posterior probability intervals. The geological time scale is in million years and follows Walker *et al.* (2009). The geographic origin of each accession is shown after the species name.

3.3. Ancestral area reconstruction

Results from the DEC analyses, using either a ML phylogram (with papaya coded as 'unknown' or as 'Central America') or an ultrametric tree obtained under the strict clock model are summarized in Table 2 and illustrated in Fig. 2. The most recent common ancestor of Caricaceae was distributed in Africa and apparently dispersed to Central America c. 35 Ma ago, where the Mexican/Guatemalan papaya clade then diversified. The family expanded from Central to South America during the Late Oligocene/Early Miocene (probably across the newly forming Isthmus; *Discussion*), eventually reaching southern South America. The widespread species *Vasconcellea cauliflora* apparently entered Central America from the south, after the formation of the

Isthmus, given that *V. cauliflora* is a rather young species; Fig. 3). By contrast, the widespread *Jacaratia spinosa*, of which we included Peruvian and Costa Rican samples, appears to have reached a large geographic range a long time ago (Fig. 3).

4. Discussion

4.1. The closest relatives of *Carica papaya* and the region of the crop's domestication

Carica papaya represents an isolated surviving lineage that diverged from its sister clade some 25 Ma ago. Wild populations of papaya are characterized by a strictly dioecious breeding system (rather than being trioecious like the cultivated papaya) and have female trees that produce small, seedy fruits with a thin mesocarp. Previous sampling of numerous papaya populations has characterized the morphological and genetic diversity of natural papayas and has found a high frequency of rare alleles among Costa Rican plants, but little differentiation among Caribbean and Pacific coastal papayas (Morshidi *et al.*, 1995; Coppens d'Eeckenbrugge *et al.*, 2007). Even higher levels of genetic diversity in the wild populations were found by Brown *et al.* (2012) who also documented the pronounced heterozygote deficiency in cultivated papaya, consistent with its history as a domesticated species. Sixteenth century Spanish explorers probably were responsible for the initial spread of papaya beyond its native Mesoamerican distribution, and 500 years of selective breeding for fruit size, shape and color, combined with selfing and inbreeding of the preferred bisexually-flowered trees (with a mal-functioning Y chromosome), probably explain the lack of genetic diversity in the cultivated papaya (Brown *et al.*, 2012). Brown *et al.* also suggested that some of today's natural papaya populations may represent descendants from papaya that were cultivated in the region of Costa Rica in the pre-Columbian era (Storey, 1976). Following the decline of pre-Columbian cultures, semi-domesticated plants could have become feral and subsequently spread throughout the region naturally, in papaya's ecological role as a pioneer species (Brown *et al.*, 2012).

An important new finding of this study is that papaya is most closely related to four species from southern Mexico and Guatemala. A morphological synapomorphy supporting this relationship is a unilocular ovary, while the remaining Caricaceae have 5-locular ovaries. Breeding efforts for papaya improvement should include the four species now revealed as the closest relatives, rather than focusing only on the highland papayas in the genus *Vasconcellea* (Sawant, 1958; Drew *et al.*, 1998; Siar *et al.*, 2011). So far, studies on the biology of the herbaceous Caricaceae are restricted to an investigation on the cultivation of the lowland species *Jarilla chocola* (Willingham and White, 1976). Other members of the papaya clade, such as *Horovitzia cnidoscoloides*, occur in montane cloud forest (Oaxaca, ca. 1250 m alt.) and might be cold adapted.

The restricted occurrence of the dioecious wild form of *C. papaya* and its four closest relatives in Central America (Storey, 1976; Moreno, 1980; Manshardt, 1992; Manshardt and Zee, 1994; Morshidi *et al.*, 1995; Badillo, 1993; Paz and Vázquez-Yanes, 1998; Coppens d'Eeckenbrugge *et al.*, 2007; Brown et al., 2012) are in line with a Central American domestication of papaya (De Candolle, 1883; Solms-Laubach, 1889; Vavilov (1940 [1992]); Storey, 1976; Manshardt, 1992; Prance and Nesbitt, 2005). Mesoamerica is one of the World's centers of plant domestication. The Olmec (1500–400 BC) and Maya (2000 BC to 900 AD) had extraordinary abilities to select plant varieties through agricultural manipulation (Pope *et al.*, 2001; Colunga-GarcíaMarín and Zizumbo-Villarreal, 2004; VanDerwarker, 2006). However, plant domestication appears to have begun in the lowland habitats of the Pacific slope of southwestern Mexico before 5000–4000 BC, greatly predating the Olmec and Mayan farming cultures (Pohl *et al.*, 1996). A phylogeographic study of *C. papaya*, covering the complete geographical range of the species and including many populations of wild and cultivated forms, would be needed to infer the direction and timing of anthropogenic range expansion.

Table 2. Inferred ancestral areas at branches (only the two most probable reconstructions are shown), and the relative probability of an area is reported for main branches. Letters in square brackets are the ranges inherited by the respective descendant branches. Values are the Log-likelihood, followed by the relative probability of the estimated reconstruction. The areas coded were: A = Africa, S = South America, C = Central America, and U = unknown.

Node	DEC — Reconstruction on a ML phylogram — Papaya coded as unknown	DEC — Reconstruction on a ML phylogram — Papaya coded as Central America	DEC — Reconstruction on a chronogram obtained under a strict clock — Papaya coded as Central America
Root: Caricaceae /	[AIA]-46.33/0.3818	[AIA]-38.79/0.4993	[ACIA]-23.9/0.4229
Moringaceae	[ACIA]-47.02/0.191	[ACIA]-39.31/0.2965	[AIA]-23.9/0.4131
Caricaceae	[CIA]-46.33/0.3816	[CIA]-38.43/0.7145	[CIA]-23.25/0.8163
Neotropics/Africa	[SIA]-46.74/0.2541	[SIA]-39.38/0.2757	[SIA]-24.74/0.1837
Neotropical clade	[CIC]-46.7/0.2636	[CIC] -38.9/0.4432	[CIC]-23.51/0.6325
Central/S. America	[CIS]-46.9/0.2164	[CIS]-39.03/0.3922	[CISC]-24.6/0.2149
Jacaratia /	[SIS]-46.02/0.5211	[SCIS]-38.61/0.5941	[SCIS]-23.25/0.817
Vasconcellea	[SCIS]-46.17/0.447	[SIS]-39.04/0.3861	[SIS]-24.99/0.1432
Jacaratia	[SIS]-46.04/0.5118	[SCIS] -38.88/0.4528	[SCIC]-24.28/0.2905
J. spinosa / *Jacaratia* spp.	[SCIS]-46.32/0.3849	[SIS] -39.1/0.3653	[SCIS]-24.28/0.2905
Vasconcellea	[SIS]-45.37/0.9959	[SIS]-38.09/0.9977	[SIS]-23.05/0.9986
Papaya clade	[UIC]-45.74/0.6873	[CIC]-38.09/0.9988	[CIC]-23.05/1
C. papaya/Mex. spp.	[CIC]-47.43/0.1271		

4.2. Origin and evolution of the Caricaceae

Based on outgroup analysis, the Caricaceae originated in Africa, where two species still occur today (Fig. 2). During the Late Cenozoic, Africa was characterized by extreme climate variability with alternating periods of high moisture levels and extreme aridity (Sepulchre *et al.*, 2006; Trauth *et al.*, 2009). A change from wet to dry conditions occurred between 4 and 3 Ma, the time when the west and east African *Cylicomorpha solmsii* and *C. parviflora* are inferred to have diverged from each other (c. 3 Ma; Table 1). Both species are big trees occurring in montane and sub-montane rainforest or along rain forest margins and paths at 500–1500 m elevation (Fig. 1 D and E). Their ranges are likely to result from the fragmentation of evergreen tropical forests during the

63

Pliocene, and their divergence time matches the inferred ages of other east African and west African rainforest clades. For example, species of *Isolona* restricted to west and central African rainforests diverged from relatives in east Africa around 4.5 Ma (Couvreur *et al.*, 2011).

The ancestral area reconstruction suggests dispersal from Africa to Central America c. 35 Ma ago (Figs. 2 and 3), possibly via a floating island carried by ocean currents from the Congo delta via the North Atlantic Equatorial Current (Houle, 1999; Fratantoni *et al.*, 2000; Renner, 2004; Antoine *et al.*, 2011). Dispersal from Africa to Central America has also been inferred for gekkonid lizards of the genus *Tarentola* and amphisbaenians (Amphisbaenidae, the Cuban genus *Cadea*), which apparently were transported on rafting vegetation from the west coast of Northwestern Africa to the West Indies (Carranza *et al.*, 2000; Vidal *et al.*, 2008). Caricaceae have soft, fleshy fruits not suitable for water dispersal, but seeds could have been transported in floating vegetation. Even if transport took several weeks, seeds might not have germinated because germination in the family is slow and erratic, due to inhibitors present in the sarcotesta (Paz and Vázquez-Yanes, 1998; Tokuhisa *et al.*, 2007).

The chronogram (Fig. 3, Table 1) in combination with the ancestral area reconstruction (Fig. 2, Table 2) implies that Caricaceae reached South America from Central America between 27 and 19 Ma ago, which matches recent geological evidence suggesting that the formation of the Isthmus of Panama already began 23–25 Ma ago, earlier than previously thought (Farris *et al.*, 2011). This may have facilitated range expansion from Mexico to Colombia, where a newly established population then began to diversify and gradually to expand the family's range south to Paraguay, Uruguay, and Argentina. The climate around 27 Ma ago was still warm and moist, prior to the Late Miocene cooling at 14 Ma (Zachos *et al.*, 2001). Mountain building in the northern Andes first peaked around 23 Ma and again around 12 Ma (Hoorn *et al.*, 2010). It was during this period that the *Vasconcellea/Jacaratia* clade started to diversify (Table 1, Fig. 3; around 19 Ma ago). Today, 18 out of 20 species of *Vasconcellea* occur in the northern Andean region, with 14 species found at altitudes between 750 and 2500 m (Scheldeman *et al.*,

2007). This supports Aradhya *et al.*'s (1999) assessment that adaptive radiation into ecologically diverse habitats during the Andean uplift led to the diversification in *Vasconcellea*. The close relationship among *V. chilensis*, *V. candicans* and *V. quercifolia* found here (Fig. 2) matches a morphological synapomorphy, namely entire to slightly pinnatifid leaves, while all other Caricaceae have deeply pinnatifid leaves. A morphological trait that would link *V. glandulosa* to this group, however, is unknown, and a better morphological characterization is needed to evaluate possible synapomorphies in *Vasconcellea*. Another group of closely related species is formed by *V. stipulata*, *V. parviflora* and *V. weberbaueri.*

Different from the diverse Andean *Vasconcellea* clade, a single species of *Jacaratia* (*J. chocoensis*) occurs in the Andean foothills. Instead, the genus *Jacaratia* appears to have adapted to the drier climates and more open vegetation that spread during the Late Miocene. Between 12 and 7 Ma ago, South America comprised large areas with tropical dry woodlands and grasslands (Pound *et al.*, 2011). This would have favored species adapted to dry, open environments or semi-deciduous forest, such as *Jacaratia corumbensis* and the baobab-like water-storing tree trunk (often well over a meter in diameter) of the Mexican *J. mexicana*. However, a more detailed study with geo-referenced specimens is needed to provide a better understanding of species distributions and habitat requirements.

5. Conclusions

Carica papaya is part of a small clade confined to Mexico and Guatemala that also includes three perennial herbs (*Jarilla chocola, J. heterophylla* and *J. nana*) and a treelet with spongy thin stems (*Horovitzia cnidoscoloides*). The geographical distribution of this clade and the occurrence wild papayas in Central America are consistent with a domestication of papaya there. The biogeographic history of Caricaceae involves long distance dispersal from Africa to Central America c. 35 Ma ago and expansion across the Panamanian land bridge sometime between 27 and 19 Ma. Diversification of *Vasconcellea,* the largest genus of the family, is related to the peak of the northern

Andean orogeny, while diversification of *Jacaratia* appears linked to the expansion of drought-adapted vegetation during the Late Miocene.

6. Acknowledgments

We thank M. Olson, C. Hughes, and an anonymous reviewer for comments that improved the manuscript. T. Kyndt for some DNA extracts, J. P. Gogue for material of *Cylicomorpha solmsii*, M. Nicholson for material of *C. parviflora*, E. Honorio for material of *Jacaratia*, and the curators of the herbaria listed in Appendix S1 for loans and authorization for removing material for DNA extraction. The first author is supported by a fellowship from Conselho Nacional de Desenvolvimento Científico e Tecnológico (CNPq), and additional funding came from the German Research Foundation (DFG RE 603/13).

Appendix A. Supplementary material
Supplementary data associated with this article can be found, in the online version, at http://dx.doi.org/10.1016/j.ympev.2012.05.019.

7. References

Antoine, P.O., Marivaux, L., Croft, D.A, Billet, G., Ganerod, M., Jaramillo, C., Martin, T., Orliac, M.J., Tejada, J., Altamirano, A.J., Duranthon, F., Fanjat, G., Rousse, S., Gismondi, R.S., 2012. Middle Eocene rodents from Peruvian Amazonia reveal the pattern and timing of caviomorph origins and biogeography. Proc. R. Soc. Lond. B Biol. Sci. 279, 1319–1326.

Aradhya, M.K., Manshardt, R.M., Zee, F., Morden, C.W., 1999. A phylogenetic analysis of the genus *Carica* L. (Caricaceae) based on restriction fragment length variation in a cpDNA intergenic spacer region. Genet. Resour. Crop Evol. 46, 579–586.

Badillo, V.M., 1971. Monografía de la família Caricaceae. Asociación de profesores, Universidad Central de Venezuela, Facultad de Agronomía, Maracay.

Badillo, V.M., 1993. Caricaceae. Segundo Esquema. Rev. Fac. Agron. Univ. Cent. Venezuela 43, 1–111.

Badillo, V.M., 2000. *Vasconcella* St.-Hil. (Caricaceae) con la rehabilitacion de este ultimo. Ernstia 10, 74–79.

Beilstein, M.A, Nagalingum, N.S., Clements, M.D., Manchester, S.R., Mathews, S., 2010. Dated molecular phylogenies indicate a Miocene origin for *Arabidopsis thaliana*. Proc. Natl. Acad. Sci. USA 107, 18724–18728.

Bell, C.D., Soltis, D.E., Soltis, P.S., 2010. The age and diversification of the angiosperms re-revisited. Am. J. Bot. 97, 1296–1303.

Brown, J.E., Bauman, J.M., Lawrie, J.F., Rocha, O.J., Moore, R.C., 2012. The structure of morphological and genetic diversity in natural populations of *Carica papaya* (Caricaceae) in Costa Rica. Biotropica 44, 179-188.

Brücher, H., 1987. The Isthmus of Panama as a crossroad for prehistoric migration of domesticated plants. GeoJournal 14, 121–122.

Brücher, H., 1989. Useful plants of Neotropical origin and their wild relatives. Springer, Berlin.

Carranza, S., Arnold, E.N., Mateo, J.A., López–Jurado, L.F., 2000. Long-distance colonization and radiation in gekkonid lizards, *Tarentola* (Reptilia: Gekkonidae), revealed by mitochondrial DNA sequences. Proc. R. Soc. Lond. B Biol. Sci. 267, 637–649.

Castresana, J., 2000. Selection of conserved blocks from multiple alignments for their use in phylogenetic analysis. Mol. Biol. Evol. 17, 540–52.

Colunga–GarcíaMarín, P., Zizumbo–Villarreal, D., 2004. Domestication of plants in Maya lowlands. Econ. Bot. 58, S101–S110.

Coppens d'Eeckenbrugge, G., Restrepo, M.T., Jiménez, D., 2007. Morphological and isozyme characterization of common papaya in Costa Rica. Acta hort. 740, 109–120.

Couvreur, T.L.P., Porter–Morgan, H., Wieringa, J.J., Chatrou, L.W., 2011. Little ecological divergence associated with speciation in two African rain forest tree genera. BMC Evol. Biol. 11, 296.

De Candolle, A., 1883. Origine des plantes cultivées. G. Baillière et cie, Paris.

Diaz-Luna, C.L., Lomeli-Sención, J.A., 1992. Revisión del género *Jarilla* Rusby (Caricaceae). Acta Bot. Mexicana 20, 77–99.

Dinesh, M., Rekha, A., Ravishankar, K., Praveen, K., Santosh, L., 2007. Breaking the intergeneric crossing barrier in papaya using sucrose treatment. Sci. Hortic. 114, 33–36.

Drew, R.A., Magdalita, P.M., O'Brien, C.M., 1998. Development of *Carica* interspecific hybrids. Acta hort. 461, 285–292.

Drummond, A.J., Rambaut, A., 2007. BEAST: Bayesian evolutionary analysis by sampling trees. BMC Evol. Biol. 7, 214.

FAOStat, 2011. Statistical databases of the Food and Agriculture Organization of the United Nations. <http://www.apps.fao.org> (Accessed on 15 October 2011).

Farris, D.W., Jaramillo, C., Bayona, G., Restrepo-Moreno, S.A., Montes, C., Cardona, A., Mora, A., Speakman, R.J., Glascock, M.D., Valencia, V., 2011. Fracturing of the Panamanian Isthmus during initial collision with South America. Geology 39, 1007–1010.

Fratantoni, D.M., Johns, W.E., Townsend, T.L., Hurlburt, H.E., 2000. Low-latitude circulation and mass transport pathways in a model of the tropical Atlantic ocean. J. Phys. Oceanogr. 30, 1944–1966.

Hoorn, C., Wesselingh, F.P., ter Steege, H., Bermudez, M.A., Mora, A., Sevink, J., Sanmartín, I., Sanchez–Meseguer, A., Anderson, C.L., Figueiredo, J.P., Jaramillo, C., Riff, D., Negri, F.R., Hooghiemstra, H., Lundberg, J., Stadler,

T., Särkinen, T., Antonelli, A., 2010. Amazonia through time: Andean uplift, climate change, landscape evolution, and biodiversity. Science 330, 927–931.

Houle, A, 1999. The origin of platyrrhines: An evaluation of the Antarctic scenario and the floating island model. Am. J. Phys. Anthropol. 109, 541–59.

Jobin-Decor, M.P., Graham, G.C., Henry, R.J., Drew, R.A., 1997. RAPD and isozyme analysis of genetic relationships between Carica papaya and wild relatives. Genet. Resour. Crop Evol. 44, 471–477.

Katoh, K., Toh, H., 2008. Recent developments in the MAFFT multiple sequence alignment program. Briefings Bioinf. 9, 286–298.

Krishnaiah, D., Awang, B., Rosalam, S., Buhri, A., 2002. Commercialization of papain enzyme from papaya, in: Omar, R., Rahman, Z.A., Latif, M.T., Lihan, T., Adam J.H. (Eds.), Proceedings of the Regional Symposium on Environment and Natural Resources, Kuala Lumpur, pp. 244–250.

Kyndt, T., Van Droogenbroeck, B., Romeijn-Peeters, E., Romero-Motochi, J.P., Scheldeman, X., Goetghebeur, P., Van Damme, P., Gheysen, G., 2005a. Molecular phylogeny and evolution of Caricaceae based on rDNA internal transcribed spacers and chloroplast sequence data. Mol. Phylogenet. Evol. 37, 442–59.

Kyndt, T., Romeijn-Peeters, E., Van Droogenbroeck, B., Romero-Motochi, J.P., Gheysen, G., Goetghebeur, P., 2005b. Species relationships in the genus Vasconcellea (Caricaceae) based on molecular and morphological evidence. Am. J. Bot. 92, 1033–1044.

Liu, Z., Moore, P.H., Ma, H., Ackerman, C.M., Ragiba, M., Yu, Q., Pearl, H.M., Kim, M.S., Charlton, J.W., Stiles, J.I., Zee, F.T., Paterson, A.H., Ming, R., 2004. A primitive Y chromosome in papaya marks incipient sex chromosome evolution. Nature 427, 348–52.

Lleras, E., 2010. Caricaceae Dumort. in Lista de Espécies da Flora do Brasil. Jardim Botânico do Rio de Janeiro. http://floradobrasil.jbrj.gov.br/2010/FB000079 (Accessed on 15 October 2011).

Lorence, D.H., Colin, R.T., 1988. Carica cnidoscoloides (sp. nov.) and sect. Holostigma (sect. nov.) of Caricaceae from Southern Mexico. Syst. Bot. 13, 107–110.

Maddison, D.R., Maddison, W.P., 2005. MacClade 4: Analysis of phylogeny and character evolution. Version 4.08a. http://macclade.org

Manshardt, R.M., 1992. Papaya, in: Hammerschlag, F.A., Litz, R.E. (Eds.), Biotechnology of perennial fruit crops, Cambridge University Press, Oxford, pp. 489–511.

Manshardt, R.M., Zee, F.T.P., 1994. Papaya germplasm and breeding in Hawaii. Fruit Var. J. 48, 146–152.

McVaugh, R., 2001. Caricaceae, in: Flora Novo-Galiciana. A descriptive account of the vascular plants of Western Mexico. Vol. 3 Ochnaceae to Losaceae, The University of Michigan Press, Ann Arbor, Michigan, pp.71–73.

Ming, R., Bendahmane, A., Renner, S.S., 2011. Sex chromosomes in land plants. Ann. Rev. Plant Bio. 62, 485–514.

Ming, R., Hou, S., Feng, Y., et al., 2008. The draft genome of the transgenic tropical fruit tree papaya (Carica papaya Linnaeus). Nature 452, 991–996.

Moreno, N.P. 1980. Caricaceae. Flora de Veracruz, Fasc. 10. Instituto de Ecología, Xalapa.

Morshidi, M., Manshardt R.M., Zee F., 1995. Isozyme variability in wild and cultivated Carica papaya. HortScience 30, 809.

O'Dea, A., Jackson, J.B.C., Fortunato, H., Smith, J.T., D'Croz, L., Johnson, K.G., Todd, J.A., 2007. Environmental change preceded Caribbean extinction by 2 million years. Proc. Natl. Acad. Sci. USA 104, 5501–5506.

Olson, M.E., 2002. Intergeneric relationships within the Caricaceae-Moringaceae clade (Brassicales) and potential morphological synapomorphies of the clade and its families. Int. J. Plant Sci. 163, 51–65.

Paz, L., Vázquez-Yanes, C. 1998. Comparative seed ecophysiology of wild and cultivated Carica papaya trees from a tropical rain forest region in Mexico. Tree Physiol. 18, 277–280.

Pohl M.D., Pope, K.O., Jones, J.G., Jacob, J.S., Piperno, D.R., deFrance, S.D., Lentz, D.L., Gifford, J.A., Danforth, M.E., Josserand, J.K. 1996. Early agriculture in the Maya lowlands. Lat. Am. Antiq. 7, 355–372.

Pope, K.O., Pohl, M.E., Jones, J.G., Lentz, D.L., von Nagy, C., Vega, F.J., Quitmyer, I.R., 2001. Origin and environmental setting of ancient agriculture in the lowlands of Mesoamerica. Science 292, 1370–1373.

Pound, M.J., Haywood, A.M., Salzmann, U., Riding, J.B., Lunt, D.J., Hunter, S.J., 2011. A Tortonian (Late Miocene, 11.61–7.25Ma) global vegetation reconstruction. Palaeogeogr Palaeoclimatol. Palaeoecol. 300, 29–45.

Prance, G.T., 1984. The pejibaye, Guilielma gasipaes (H.B.K.) Bailey, and the papaya, Carica papaya L. in: D. Stone (Ed.), Pre-Columbian Plant Migration. Harvard University Press, Massachusetts, pp. 85–104.

Prance, G.T., Nesbitt, M. 2005. The Cultural History of Plants, first ed. Routledge, London/New York.

Ree, R.H., Moore, B.R., Webb, C.O., Donoghue, M.J., 2005. A Likelihood Framework for Inferring the Evolution of Geographic Range on Phylogenetic Trees. Evolution 59, 2299–2311.

Ree, R.H., Smith, S.A., 2008. Maximum likelihood inference of geographic range evolution by dispersal, local extinction, and cladogenesis. Syst. Biol. 57, 4–14.

Renner, S., 2004. Plant dispersal across the tropical Atlantic by wind and sea currents. Int. J. Plant Sci. 165, S23–S33.

Sawant, A.C., 1958. Crossing relationships in the genus Carica. Evolution 12, 263–266.

Scheldeman, X., Willemen, L., Coppens d'Eeckenbrugge, G., Romeijn–Peeters, E., Restrepo, M.T., Romero Motoche, J., Jiménez, D., Lobo, M., Medina, C.I., Reyes, C., Rodríguez, D., Ocampo, J. A., Damme, P., Goetgebeur, P., 2007. Distribution, diversity and environmental adaptation of highland papayas (Vasconcellea spp.) in tropical and subtropical America. Biodivers. Conserv. 16, 1867–1884.

Scheldeman, X., Kyndt, T., Coppens d'Eeckenbrugge, G.C., Ming, R., Drew, R., Van Droogenbroeck, B.V., Van Damme, P., Moore, P.H., 2011. *Vasconcellea*. In: Kole, C. (Ed.), Wild crop relatives: genomic and breeding resources. Tropical and subtropical fruits, first ed. Springer-Verlag, Berlin Heidelberg.

Sebastian, P., Schaefer, H., Telford, I.R.H., Renner, S.S., 2010. Cucumber (*Cucumis sativus*) and melon (*C. melo*) have numerous wild relatives in Asia and Australia, and the sister species of melon is from Australia. Proc. Natl. Acad. Sci. USA 107, 14269–14273.

Sepulchre, P., Ramstein, G., Fluteau, F., Schuster, M., Tiercelin, J.J., Brunet, M., 2006. Tectonic uplift and Eastern Africa aridification. Science 313, 1419–1423.

Siar, S.V., Beligan, G.A., Sajise, A.J.C., Villegas, V.N., Drew, R.A, 2011. Papaya ringspot virus resistance in *Carica papaya* via introgression from *Vasconcellea quercifolia*. Euphytica 181, 159–168.

Silvestro, D., Michalak, I., 2011. raxmlGUI: a graphical front-end for RAxML. Org. Divers. Evol., doi 10.1007/s13127-011-0056-0

Solms–Laubach, 1889. Die Heimat und der Ursprung des kultivierten Melonenbaumes, *Carica papaya* L. Bot. Ztg. 44, 709–720.

Stamatakis, A., 2006. RAxML–VI–HPC: maximum likelihood–based phylogenetic analyses with thousands of taxa and mixed models. Bioinformatics 22, 2688–2690.

Storey, W.B., 1976. Papaya, *Carica papaya*. in: Simmonds, N.W. (Ed.), Evolution of Crop Plants, Longman, London, pp. 21–24.

Tokuhisa, D., Cunha, D., Dos, F., Dias, S., Alvarenga, E.M., Hilst, P.C., Demuner, A.J., 2007. Compostos fenólicos inibidores da germinação de sementes de mamão (*Carica papaya* L.). Rev. Bras. Sementes 29, 161–168.

Trauth, M.H., Larrasoaña, J.C., Mudelsee, M., 2009. Trends, rhythms and events in Plio-Pleistocene African climate. Quat. Sci. Rev. 28, 399–411.

VanDerwarker, A.M. 2006. Farming, Hunting, and Fishing in the Olmec World, first ed. The University of Texas Press, Austin.

Van Droogenbroeck, B.V., Breyne, P., Goetghebeur, P., Romeijn–Peeters, E., Kyndt, T., Gheysen, G., 2002. AFLP analysis of genetic relationships among papaya and its wild relatives (Caricaceae) from Ecuador. Theor. Appl. Genet. 105, 289–297.

Van Droogenbroeck, B.V., Kyndt, T., Maertens, I., Romeijn–Peeters, E., Scheldeman, X., Romero–Motochi, J.P., Van Damme, P., Goetghebeur, P., Gheysen, G., 2004. Phylogenetic analysis of the highland papayas (*Vasconcellea*) and allied genera (Caricaceae) using PCR–RFLP. Theor. Appl. Genet. 108, 1473–1486.

Vavilov, N.I., 1940 [1992 Engl. translation] Origin and Geography of Cultivated Plants, Translated by D. Löve. Cambridge University Press, Cambridge.

Vidal, N., Azvolinsky, A., Cruaud, C., Hedges, S.B., 2008. Origin of tropical American burrowing reptiles by transatlantic rafting. Biol. Lett. 4, 115–118.

Walker, J.D., Geissman, J.W., compilers, 2009. Geologic Time Scale: Geological Society of America, doi: 10.1130/2009.

Wikström, N., Savolainen, V., Chase, M.W., 2001. Evolution of the angiosperms: calibrating the family tree. Proc. Roy. Soc. Lond. B 268, 2211–2220.

Willingham, B.C., White, G.A., 1976. Agronomic evaluation of prospective new crop species V. *Jarilla chocola*: A proteinase source. Econ. Bot. 30, 189–192.

Wu, X., Wang, J., Na, J.K., Yu, Q., Moore, R.C., Zee, F., Huber, S.C., Ming, R., 2010. The origin of the non-recombining region of sex chromosomes in *Carica* and *Vasconcellea*. Plant J. 63, 801–810.

Zachos, J., Pagani, M., Sloan, L., Thomas, E., Billups, K., 2001. Trends, rhythms, and aberrations in global climate 65 Ma to present. Science 292, 686–693.

Figure S1. A Bayesian tree obtained from the same data as Fig. 2. Values on nodes are the posterior probabilities.

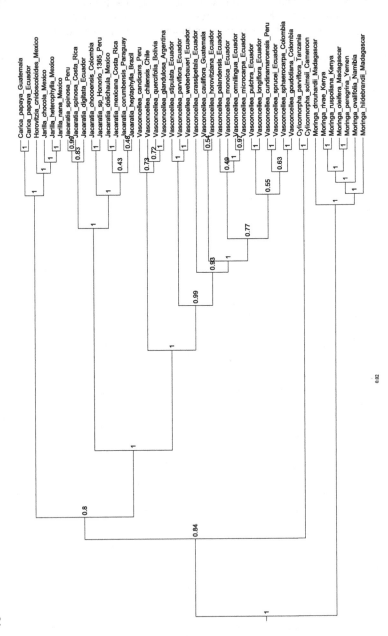

Appendix S1. Species names and their authors, herbarium vouchers, geographic provenience, and GenBank accession numbers. Herbarium acronyms follow the Index Herbariorum at http://sciweb.nybg.org/science2/IndexHerbariorum.asp. Sequences in bold were used to produce the complete phylogeny of the family (Fig. 2). Species included in the dating analyses are marked with an asterisk before the name. Sequences newly generated for this study have GenBank numbers beginning with JX. The other sequences were produced by Kyndt et al. (2005a).

Species	Taxon distribution and habitat	Herbarium vouchers and their geographic origin	*rbc*L	*mat*K	*Trn*L-*trn*F	*psb*A-*trn*H	*rpl*20-*rps*12	ITS region
Cylicomorpha parviflora Urb.	Tropical east Africa (Kenya, Tanzania and Malawi). Rain forest	Mwangoka, M.A. 3212, Tanzania (M)	JX091915	JX092004	JX091824	JX091964	JX091876	JX092052
Cylicomorpha solmsii (Urb.) Urb.	Tropical west Africa (Nigeria, Cameroon, Congo and Central African Republic). Rain Forest	Ghogue, J.P. 2115, Yaounde, Eloumden 3°49'25"N, 11°26'27"E Alt: 900 m (YA)	JX091916	JX092005	JX091825	JX091965	JX091877	JX092053
Carica papaya L.	Neotropics and naturalized in nearly all tropical regions of the Old World	Kufer, J. 399, Guatemala 14°49'01"N, 89°23'21"E Alt.: 450 m (M)	JX091913	JX092002	JX091823	JX091963	JX091875	JX092051
		Romeijn-Peeters, E.H. 57, Ecuador, Catacocha, Loja (GENT)	JX091914	JX092003 AY461576	DQ61124	AY847053	JX091874	AY461564
Horovitzia cnidoscoloides (Lorence & Torres) Badillo	Mexico (Oaxaca). Cloud forests	Torres, R.C. 8167, Mexico, Oaxaca, Ixtlán, Sierra de Juaréz. 17°19'35"N, 96°28'41"W. Alt.: 1250 m (MEXU)	JX091917	JX092006	JX091826	JX091966	JX091878	JX092054
Jarilla chocola Standl.	Mexico (Sonora, Chihuahua, Sinaloa, Nayarit, Jalisco, Michoacan, Chiapas) and Guatemala (Jutiapa). Humid deciduous forests of the Pacific slopes	Lott, E.J. 31, Mexico, Jalisco, San Patricio, Estación de Biología Chamela. 19°39'13"N, 104°51'26"W (MEXU)	JX091927	JX092017	JX091838	JX091977	JX091884	JX092064
		Gentry, H.S. 1553, Mexico 31 (F)	-	-	JX091837	JX091976	-	

Species	Taxon distribution and habitat	Herbarium vouchers and their geographic origin	rbcL	matK	TrnL-trnF	psbA-trnH	rpl20-rps12	ITS region
*Jarilla heterophylla (Llave) Rusby	Mexico (Baja California Sur, Jalisco, Michoacán, Guanajuato). Deciduous forest	Lomelí, J.A. 20003, Mexico, Jalisco, Zacoalco. 20°13'35"N, 103°35'26"W (F)	JX091926	JX092016	JX091839	JX091975	JX091885	JX092065
*Jarilla nana (Benth.) McVaugh	Mexico (Zacatecas, Jalisco, Guanajuto, Michoacan, Hidalgo, Mexico, Distrito Federal). Dry tropical deciduous forest or Oak forest	Lomelí, J.A. 20002, Mexico, Jalisco, Chiquilistlán. 20°03'66"N 10°54'23"W (F)	JX091928	JX092018	JX091840	JX091978	JX091886	JX092066
Jacaratia chocoensis A.H. Gentry & Forero	Colombia (Chocó). Rain forest	Fonnegra, R. 5758, Colombia, Antioquia, San Luis. 6°02'03"N, 75°01'38"W Alt.: 1300–1500 m (MO)	-	JX092007	JX091828	JX091968	-	JX092055
Jacaratia corumbensis Kuntze	Southeastern Bolivia, Paraguay, northern Argentina, and small area of SW Brazil. Xerophytic sandy areas	Fiebrig, K. 1468, Paraguay (M)	JX091918	JX092008	JX091829	JX091969	JX091879	JX092056
Jacaratia digitata (Poepp. & Endl.) Solms	West of the Amazon Basin (Colombia, Ecuador, Peru, Bolivia and Brazil). Rain forest	Romeijn-Peeters, E.H. 36, Zamora, Ecuador (GENT)	JX091919	JX092009 AY461574	JX091830	-	JX091880	-
		Monetagudo, A. 19254, Ecuador, Bogi. Alt.: 270 m (LOJA)	-	-	JX091831	-	-	JX092057
*Jacaratia dolichaula (Donn. Sm.) Woodson	Central America (from South Mexico to Panama). Semideciduous forests	Calzada, J.I. 4785, Mexico, Veracruz, San Andrés Tuxtla. 18°35'N, 95°04'W (F)	JX091920	JX092010	JX091832	JX091970	JX091881	JX092058
Jacaratia heptaphylla (Vell.) A. DC.	Brazil (Bahia, Minas Gerais, Espírito Santo, Rio de Janeiro, Mato Grosso do Sul, and São Paulo). Wet Forests	Oliveira, P.P. 670, Brasil, Rio de Janeiro, Rio das Ostras. 22°25'38"S, 42°2'8.5"W (BHCB)	JX091921	JX092011	JX091833	JX091973	-	JX092059

Species	Taxon distribution and habitat	Herbarium vouchers and their geographic origin	rbcL	matK	TrnL-trnF	psbA-trnH	rpl20-rps12	ITS region
*Jacaratia mexicana A. DC.	Mexico, El Salvador and Nicaragua. Floodplains, slopes in semi-deciduous forests	Léon, J. 2427, Costa Rica, Cartago, Turrialba (M)	JX091922	JX092012	JX091834	JX091971	-	JX092060
*Jacaratia spinosa (Aubl.) A.DC.	Nicaragua, Costa Rica, Panama, Guiana, Suriname, French Guiana,	Fondur 12881 (M) Costa Rica, Cartago, Tucurrique	JX091924	JX092014	JX091835	JX091967	-	JX092061
	Ecuador, Peru, Chile, Bolivia, Argentina, Paraguay and Brazil. Wet forests, sometimes with strong dry season	Honorio, E. 1348, Peru, Cerro de Pasco, Oxapampa, 10°10'23.1"S 75°34'20.3"W Alt.: 1057 (M)	JX091925	JX092015	JX091836	JX091972	JX091883	JX092062
*Jacaratia sp	Possible new species known only from two Peruvian collections. Rain forest	Honorio, E. 1365, Peru, Cerro de Pasco, Oxapampa, 10°10'55.7"S 75°34'24.8"W (M)	JX091923	JX092013	JX091827	JX091974	JX091882	JX092063
*Vasconcellea candicans (A. Gray) A. DC.	Peru. Relatively dry areas or rocky outcrops	Leiva, L. 1201, Peru, La Liberdad, Otuzco (MO)	JX091936	JX092025	JX091848	JX091986	JX091892	JX092074
		Romeijn-Peeters, E.H. 113, Catacocha, Loja, Ecuador (GENT)	JX091937	JX092024	-	-	-	-
*Vasconcellea cauliflora (Jacq.) A. DC.	From southern Mexico to northern South America (Colombia and Venezuela). Wet forests	Standley, P.C. 89272, Escuintla, Guatemala, Alt.: 700 m (F)	JX091939	JX092028	JX091850	JX091987	JX091894	JX092075
		Romeijn-Peeters, E.H. 284 (GENT)	JX091938	JX092026	JX091849	-	JX091893	-
		Smith, H.C. 838, Colombia (F)	-	JX092027	-	JX091988	JX091895	-
*Vasconcellea chilensis Planch. ex A. DC.	Chile (Coquimbo, Valparaiso) Costal areas from 0 to 2000 m	Frömbling, Chile (M)	JX091941	JX092030	JX091852	JX091990	-	JX092076
		Dessauer, s.n., Chile (M)	JX091940	JX092029	JX091851	JX091989	-	-
		Jiles, C. 219, Chile (M)	-	JX092031	-	-	-	-
		Mickel, J. s.n. (M)	-	-	JX091853	-	-	-
Vasconcellea crassipetala (V.M. Badillo) V.M. Badillo	Colombia and Ecuador Wet forests (Andes)	Romeijn-Peeters, E.H. 282 Ecuador (GENT)	JX091942	JX092032 AY461559	DQ061132	AY847039	JX091896	AY46153

Species	Taxon distribution and habitat	Herbarium vouchers and their geographic origin	rbcL	matK	TrnL-trnF	psbA-trnH	rpl20-rps12	ITS region
Vasconcellea cundinamarcensis V.M. Badillo (= Vasconcellea pubescens A.DC.)	Panama, Colombia, Venezuela, Ecuador, Peru and Bolivia. Montane forests (1500–3000 m)	Förther, H. s.n., Peru, (M)	JX091955	JX092044	JX091865	JX091996	JX091906	JX092082
		cultivated in the USDA national papaya collection in Hilo, HCAR46 (M)	JX091956	JX092045	JX091866	JX091997	JX091907	-
*Vasconcellea glandulosa A. DC..	Peru, Bolivia, Argentina and Brasil (Eastern Andes). Tropical and subtropical wet forests	Novara, L.J. 8655, Argentina (M)	JX091943	JX092033	JX091854	JX091991	JX091897	JX092077
*Vasconcellea goudotiana Triana & Planch.	Colombia (Antioquia, Boyaca, Quindio, Tolima, Cauca, Huila) and Panama (Canal zone). Subtropical forests (1500–2200 m)	Romeijn-Peeters, E.H. 285, Plant grown from seed obtained from Drew R., originally collected in Colombia (GENT)	JX091945	JX092035	JX091855 DQ061135	AY847035	JX091899	AY461540
		cultivated in the USDA national papaya collection in Hilo, HCAR167 (M)	JX091944	JX092034	JX091856	JX091992	JX091898	JX092078
Vasconcellea horovitziana (V.M. Badillo) V.M. Badillo	Eastern Ecuador (Manabi and Pichincha). Wet forests	Provided by R. Ming Originally collected in Ecuador (M)	-	AY461566	DQ061141	AY847036	-	AY461543 JX092080
Vasconcellea x heilbornii	Ecuador and Peru (1600-2800 m)	Romeijn-Peeters, E.H. 198, Ayora, Pichincha, Ecuador (GENT)	JX091947	JX092037	-	-	-	-
		Romeijn-Peeters, E.H. 155, Loja, Ecuador (GENT)	-	-	JX091857	-	-	-
Vasconcellea longiflora (V.M. Badillo) V.M. Badillo	Colombia (Antioquía, and Cauca). Subtropical wet forests (1000-2000 m)	Romeijn-Peeters, E.H. 228, Ecuador (GENT)	-	AY461557	DQ061131	AY847037	-	AY461542
Vasconcellea microcarpa (Jacq.) A. DC.	Panama, French Guiana, Venezuela, Colombia, Ecuador, Peru and Brazil. Wet forests	Romeijn-Peeters, E.H. 225, Mera, Pastaza, Ecuador (GENT)	JX091948	AY461563	JX091858 DQ061130	AY847052	-	AY461536

76

Species	Taxon distribution and habitat	Herbarium vouchers and their geographic origin	rbcL	matK	TrnL-trnF	psbA-trnH	rpl20-rps12	ITS region
*Vasconcellea monoica (Desf.) A. DC.	Ecuador, Peru and Bolivia. Wet forests	Romeijn-Peeters, E.H. 58 Ecuador (GENT)	JX091950	JX092039	JX091859 DQ61119	AY847032	JX091901	AY461537
		cultivated in the USDA national papaya collection in Hilo, HCAR171 (M)	JX091949	JX092038	JX091860	JX091994	JX091900	JX092081
*Vasconcellea omnilingua (V.M. Badillo) V.M. Badillo	Ecuador (El Oro). Wet forests (2100–2300 m)	Romeijn-Peeters, E.H. 238, Güishagüiña, El Oro, Ecuador (GENT)	JX091951	JX092040	JX091861 DQ06112	AY847042	JX091902	AY461534
*Vasconcellea palandensis (V.M. Badillo, Van den Eynden & Van Damme) V.M. Badillo	Ecuador (Palanda). Wet forests (1700–1900 m)	Romeijn-Peeters, E.H. 66, Palanda, Zamora,Ecuador (GENT)	JX091952	JX092041	JX091862 DQ061140	AY847047	JX091903	AY461535
*Vasconcellea parviflora A. DC.	Ecuador and Peru (Western Andes). Dry areas	Romeijn-Peeters, E.H. 45, Catacocha, Loja, Ecuador (GENT)	JX091954	JX092043	JX091863 DQ061122	AY847048	JX091905	AY461526
		cultivated in the USDA national papaya collection in Hilo, HCAR179(M)	JX091953	JX092042	JX091864	JX091995	JX091904	-
*Vasconcellea pulchra (V.M. Badillo) V.M. Badillo	Border of forests (1300–1400m)	Romeijn-Peeters, E.H. 191, La Con-cordia, Pichincha, Ecuador (GENT)	-	AY461567	DQ061128	AY847046	-	AY461541
*Vasconcellea quercifolia A. St.-Hil.	Southern Peru, Bolivia, Northern Argentina, Paraguay, Brazil. From near sea level in Brazil up to 3000m in the Andes	Feuerer, T. s.n., Bolivia (M)	JX091957	JX092046	JX091868	JX091998	JX091909	JX092083
		cultivated in the USDA national papaya collection in Hilo, originally collected in Paraguay HCAR226(M)	JX091958	JX092047	JX091869	JX091999	JX091910	JX092084
		Cultivated at Giessen Botanical Garden (M)	-	-	JX091867	-	JX091908	-

77

Species	Taxon distribution and habitat	Herbarium vouchers and their geographic origin	rbcL	matK	TrnL-trnF	psbA-trnH	rpl20-rps12	ITS region
Vasconcellea sphaerocarpa (Garcia & Hernandez) V.M. Badillo	Colombia (both sides of the Andes; Santander, Cundinamarca, Valle del Cauca and Huila). Rain forests (700-2440 m).	Silverstone, P. 6786, Colombia (MO)	JX091946	JX092036	JX091871	JX091993	-	JX092079
		Manshardt, R. 1054, Colombia, cultivated in the USDA national papaya collection in Hilo, HCAR284	JX091959	JX092048	JX091870	JX092000	JX091911	-
Vasconcellea sprucei (V.M. Badillo) V.M. Badillo	Ecuador (Tungurahua, Napo) Border of forests (1300–2400m)	Asplund, E. 8784, Ecuador (NY)	JX091960	-	JX091872	JX092001	-	JX092085
Vasconcellea stipulata (V.M. Badillo) V.M. Badillo	Ecuador (Azauy and Loja) and Peru (Cajamarca) Mountain forests (1600–1500 m)	Romeijn-Peeters, E.H. 55, Catacocha, Loja, Ecuador (GENT)	JX091961	JX092049 AY46157	JX091873 DQ061123	AY847051	JX091912	AY461548
Vasconcellea weberbaueri (Harms) V.M. Badillo	Peru (Amazonas, Cajamarca and La Liberdad). Andean subtropical wet forests	Romeijn-Peeters, E.H. 10, Podocarpus, Loja, Ecuador (GENT)	JX091962	JX092050 AY461573	DQ061121	-	-	AY461527
Moringa drouhardii Jum.	Southern Madagascar. Dry forest on limestone cliffs	Carvalho, F.A. 2229, cultivated at Munich Bot. Gard. (v/0451) Madagascar (M)	JX091929	JX092019	JX091841	JX091979	JX091887	JX092067
Moringa hildebrandtii Engl.	Madagascar (extinct in the wild but preserved by indigenous horticultural practices)	Carvalho, F.A. 2228, cultivated at Munich Bot. Gard. (2000/1940) Madagascar (M)	JX091930	JX092020	JX091842	JX091980	JX091888	JX092068
*Moringa oleifera Lam.	Only native in India	Carvalho, F.A. 2227, cultivated at Munich Bot. Gard. (08/1933) Unknown origin (M)	JX091931	JX092021	JX091843	JX091981	JX091889	JX092069
*Moringa ovalifolia Dinter & A. Berger	From central Namibia to southwestern Angola. Usually on very rocky ground	Carvalho, F.A. 2228, cultivated at Munich Bot. Gard. (v/0450) Namibia (M)	JX091932	JX092022	JX091844	JX091982	JX091890	JX092070

Species	Taxon distribution and habitat	Herbarium vouchers and their geographic origin	rbcL	matK	TrnL-trnF	psbA-trnH	rpl20-rps12	ITS region
*Moringa peregrina Forssk. ex Fiori	Red Sea region. Dry areas	Carvalho, F.A. 2230, cultivated at Munich Bot. Gard. (1987/1019) Jemen (M)	JX091933	JX092023	JX091845	JX091983	JX091891	JX092071
Moringa rivae Chiov.	Southern Lake Turkana to Mandera District in Kenya and southeastern Ethiopia	Olson, M. 677, Kenya (MO)	JX091934	-	JX091846	JX091984	-	JX092072
Moringa ruspoliana Engler	Northeastern Kenya, northern Somalia and southeastern Ethiopia	Olson, M. 702, Kenya (MO)	JX091935	-	JX091847	JX091985	-	JX092073

79

Appendix S2. Primer sequences used in this study (listed 5′→ 3′)

Gene or spacer region	Primer sequence (reference)
***psbA*-*trn*H**	
*psb*A	GTTATGCATGAACGTAATGCTC (Sang *et al.* 1997)
*trn*H	CGCGCATGGTGGATTCACAAATC (Sang *et al.* 1997)
***trn*L-F**	
c	CGA AAT CGG TAG ACG CTA CG (Taberlet *et al.* 1991)
d	GGG GAT AGA GGG ACT TGA AC (Taberlet *et al.* 1991)
e	GGT TCA AGT CCC TCT ATC CC (Taberlet *et al.* 1991)
f	ATT TGA ACT GGT GAC ACG AG (Taberlet *et al.* 1991)
***rbc*L**	
1f	ATGTCACCACAAACAGAAACTAAAGC (Chase *et al.* 1993)
600f	ATT TAT GCG TTG GAG AGA CCG (Kocyan *et al.* 2007)
800r	CAA TAA CRG CAT GCA TYG CAC GRT (Kocyan *et al.* 2007)
1460r	CTTTTAGTAAAAGATTGGGCCGAG (Chase *et al.* 1993)
***mat*K**	
Af	CAT TAT CCA CTT ATC TTT CAG GAG T (Ooi *et al.* 1995)
F1	GGT TTG CAC TCA TTG TGG AAA TTC C (Yokoyama *et al.* 2000)
F2	TCC TAT ATA ATT CTC ATG TAT GTG A (Yokoyama *et al.* 2000)
8r	AAA GTT CTA GCA CAA GAA AGT CGA (Yokoyama *et al.* 2000)
R1	TAC CAC TGA AGG ATT TAG TCG CAC A (Yokoyama *et al.* 2000)
R2	AAG ATG TTA ATC GTA AAT GAG AAG (Yokoyama *et al.* 2000)
ITS region	
1	TCCGTAGGTGAACCTGCGG (White *et al.* 1990)
2	GCTGCGTTCTTCATCGATGC (White *et al.* 1990)
3	GCATCGATGAAGAACGCAGC (White *et al.* 1990)
4	TCCTCCGCTTATTGATATGC (White *et al.* 1990)
Rps12-rpl20	
Rps12	GTC GAG GAA CAT GTA CTA GG (Hamilton 1999)
Rpl20	TTT GTT CTA CGT CTT CGA GC (Hamilton 1999)

References (Appendix S2)

Chase, M.W., Soltis, D.E., Olmstead, R.G. et al. 1993. Phylogenetics of Seed Plants: An Analysis of Nucleotide Sequences from the Plastid Gene rbcL. Annals of the Missouri Botanical Garden. 80, 528.

Hamilton, M.B. 1999. Four primer pairs for the amplification of chloroplast intergenic regions with intraspecific variation. Molecular Ecology. 8, 521-523.

Kocyan, A., Zhang, L.-B., Schaefer, H. & Renner, S. S. 2007. A multi-locus chloroplast phylogeny for the Cucurbitaceae and its implications for character evolution and classification. Mol. Phylogenet. Evol., 44(2), 553-77.

Ooi, K., Endo, Y., Yokoyama, J. & Murakami, N. 1995. Useful primer designs to amplify DNA fragments of the plastid gene matK from angiosperm plants. Journal of Japanese Botany. 70: 328-331.

Sang, T., Crawford, D. J. & Stuessy, T. F. 1997. Chloroplast DNA phylogeny, reticulate evolution, and biogeography of Paeonia (Paeoniaceae). American Journal of Botany. 84, 1120-1136.

Taberlet, P., Gielly, L., Pautou, G. & Bouvet, J. 1991. Universal primers for amplification of three non-coding regions of chloroplast DNA. Plant molecular biology. 17, 1105-9.

White, T.J., Bruns, T., Lee, S. & Taylor, J. 1990. Amplification and direct sequencing of fungal ribosomal RNA genes for phylogenetics. PCR Protocols: a guide to methods and applications. Innis, M.A., Gelfand, D.H., Sninsky, J.J., White, T.J. (eds.), pp. 315-322. Academic Press, New York, USA

Yokoyama, J., Suzuki, M., Iwatsuki, K. & Hasebe, M. 2000. Molecular phylogeny of Coriaria, with special emphasis on the disjunct distribution. Mol. Phylogenet. Evol. 14, 11-19.

V. Chromosome Counts for the Caricaceae Reveal Unexpected Dysploidy*

*manuscript in preparation coauthored by Fernanda Antunes Carvalho, Alexander Rockinger, Aretuza Sousa, and Susanne S. Renner

Introduction

The family Caricaceae has six genera with together 34 species and one hybrid (Carvalho 2013). Besides the economically important crop, *Carica papaya*, other species in the family also produce edible fruits that are sold in local markets. The sister group of *C. papaya*, which originated in Central America, consists of a clade of four species, three herbs in the genus *Jarilla* endemic to Mexico and Guatemala, and the single species of *Horovitzia*, *H. cnidoscoloides*, endemic to cloud forests of Sierra de Juarez in Oaxaca, southern Mexico (Carvalho and Renner 2012). This Central American papaya clade in turn is sister to the mostly Andean *Vasconcellea* and *Jacaratia* group (Fig. 1). Sister to the entire Neotropical clade is the African genus *Cylicomorpha*, which consist of two species distributed in pre-montane forests in East and West Africa.

Despite the economic importance of the Caricaceae, only 10 of their 34 species from three genera have had their chromosomes counted. Heilborn (1921) reported $2n = 18$ for *Carica papaya*, *Vasconcellea pubescens*, and the hybrid *Vasconcelllea × heilbornii*. The same number was reported for *Jacaratia spinosa* (Kumar and Srinivasan 1944; Silva et al. 2012), *Vasconcellea goudotiana*, *V. microcarpa*, *V. monoica* (de Zerpa, 1959), and *V. quercifolia* (Bernardello et al. 1990). More recent studies confirmed $2n = 18$ for these species (Costa et al. 2008; Damasceno et al. 2009; Silva et al. 2012) and reported the same number for four additional species of *Vasconcellea* (*V. cauliflora, V. longiflora, V. palandensis, V. sphaerocarpa;* Caetano et al. 2008; Costa et al. 2008; Damasceno et al. 2009; Silva et al. 2012). The genera *Jarilla*, *Horovitzia*, and *Cylicomorpha* have never had their chromosomes counted. Chromosomes in Caricaceae are relatively small (1–4.25 μm), and the chromosomes pairs can not be distinguished morphologically (Datta 1971; Damasceno et al. 2009). A study by Costa et al. (2008) provides the only molecular-cytogenetic data for Caricaceae so far: The number and position of 18S and 5S ribosomal DNA (rDNA) fluorescent *in situ* hybridization (FISH) signals in *Carica papaya, Vasconcellea pubescens,* and *V. goudotiana* varied, with the two species of *Vasconcellea* being most similar to each other.

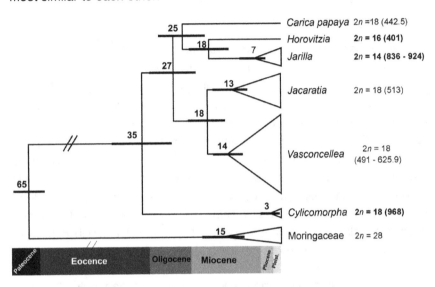

Fig. 1. Evolutionary relationships among the genera of Caricaceae, with branch lengths proportional to time and values at nodes indicating divergence times (modified from Carvalho and Renner 2012). Values in brackets refer to genome size ranges in millions of base pairs (Mbp) per haploid genome (see Table 1 for details). The chromosome number of *Moringa oleifera* is from (Silva et al. 2011). In bold are chromosome numbers and 1C-values first reported here.

Based on wild-collected material brought into cultivation in the greenhouses of the Munich Botanical Garden, we here report chromosomes numbers for *Cylicomorpha parviflora, Horovitzia cnidoscoloides, Jarilla caudata* and *J. heterophylla*, which have pivotal positions in the family as, respectively, a member of the sister genus to all Neotropical Caricaceae and the sister clade to papaya itself (Fig. 1). We also summarize all C-value measurements so far published for the

84

Caricaceae, including recently obtained measurements for the four newly counted species. These values are important for calculating the expected coverage in whole-genome sequencing, while chromosome numbers are important to determine expected linkage groups.

Material and Methods

Plant material and pretreatment – Wild-collected seeds of *Cylicomorpha parviflora, Jarilla caudata, J. heterophylla,* and *Horovitzia cnidoscoloides* were germinated, and since April 2013, seedlings have been growing in the greenhouses of the Munich Botanical Garden. Vouchers have been deposited in the Botanische Staatsammlung (M) and are listed in Table 1. Root tips were collected between 10:45 and 12:00 am and pretreated with 2 mM 8-hydroxyquinoline (8HQ). Roots of *Cylicomorpha* were kept for 20 h at 4°C, while roots of three individuals of *Horovitzia* and *Jarilla* were first kept for 3 h at room temperature and then for an additional 3 h at 4°C. Root tips of both species were then fixed in freshly prepared ethanol: acetic acid (3:1) overnight at room temperature and stored at –20°C.

Chromosomes preparation — The fixed roots tips were washed with dH_2O in three baths of 5 min each, and subsequently digested with 10 μl of an enzyme mix (0.4% pectolyase, 0.4% cytohelicase, 1% cellulase in citrate buffer) during 5 min at 37°C. After that, the excess of enzyme was removed with a filter paper, and roots were washed and incubated in dH2O for 30 min. Under a binocular, root meristems were dissected in 45% acetic acid, squashed, and covered with coverslips. The quality of spreads was checked microscopically using phase-contrast, and the best slides were selected for further analysis. The slides were dried on a cold plate at –40°C during 30 min, and then coverslips were removed for further drying at room temperature.

DAPI staining and visualization — Chromosomes were counterstained with 10 μl of diamidino-2-phenylindol (DAPI, 2 $\mu g/ml$) and mounted in Vectashield (Vector Laboratories, Burlingame, California, USA). The slides were kept dark at room temperature during at least 1 h. Images were taken with a Leica DMR microscope equipped with a KAPPA-CCD camera, and the KAPPA software.

Results and Discussion

The African species *Cylicomorpha parviflora* has $2n = 18$ small chromosomes of homogeneous size (a count based on 22 metaphases; Fig. 2D). This number was also reported from 11 species in four of the family's six genera (Table 1). Unexpectedly, the closest relatives of *C. papaya* do not share that number. Instead, *Horovitzia cnidoscoloides* has $2n = 16$ (based on 9 metaphases; Fig. 2C), and the two species of *Jarilla* have $2n = 14$ (based on 6 metaphases for *J. heterophylla* and 7 metaphases for *Jarilla caudata*; Figs. 2A and B). Genome sizes so far known in Caricaceae are summarized in Table 1. The genome size of *Cylicomorpha parviflora* is about 968 Mb per haploid genome, and is much larger than that of any other Caricaceae species (Table 1). The two *Jarilla* species also have relatively large genomes, being 924 Mbp in *Jarilla caudata* and 836 Mbp in *Jarilla heterophylla.* The genome size of *H. cnidoscoloides* is 401 Mbp, similar to the genome size of *Carica papaya* (442.5 Mbp; Gschwend et al. 2013).

Based on the available counts, polyploidy plays no role in the Caricaceae. Instead, there is a dysploid reduction in chromosome number that must have begun in the most recent common ancestor of *Horovitzia* and *Jarilla* (Fig. 1), possibly involving reduction a fusion of two chromosomes, which would explain the change from $n = 9$ to $n = 8$. Dysploid reductions in chromosome number have been analyzed in detail in *Arabidopsis* (Yogeeswaran et al. 2005; Lysak et al. 2006; Mandakova and Lysak 2008), Triticeae (Luo et al. 2009), and recently *Cucumis* in fully sequenced genomes of cucumber and its sister species were compared (Yang et al. 2013). In *Arabidopsis*, gradual rearrangements involving inversions, fusions, and translocations led from an ancestral $n = 8$ to $n = 5$ in *A. thaliana* (Lysak et al. 2006).

86

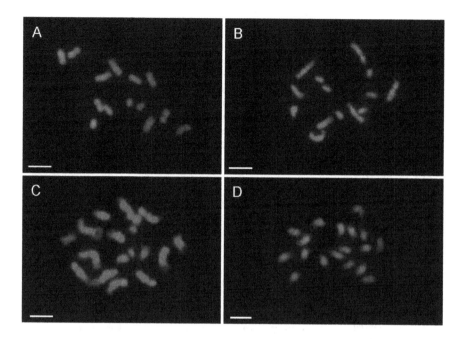

Fig. 2. DAPI-stained metaphase chromosomes. **A**, *Jarilla caudata* (2*n* = 14); **B**, *Jarilla heterophylla* (2*n* = 14); **C**, *Horovitzia cnidoscoloides* (2*n* = 16); and **D**, *Cylicomorpha parviflora* (2*n* = 18). Bar corresponds to 5 μm

In *Cucumis sativus*, a similar mix of mechanisms led to dysploid chromosome reduction from an *n* = 12 ancestor to the *n* = 7 karyotype of cucumber (Yang et al. 2013). Different from the other genera so far studied, large and small chromosome pairs are identifiable in the karyotype the two *Jarilla* species (Fig. 2A, B). This finding supports our interpretation of the *Jarilla* karyotype having originated from a past fusion event in an ancestral karyotype with *n* = 9. Genomic and cytogenetic analyses of entire syntenic genomes in principle allow inferring the details of such past rearrangements, and the present results make the Caricaceae another family in which the mechanisms of dysploidy could fruitfully be studied.

Table 1. Chromosome counts and genome size for species of Caricaceae with references, including results first reported here. Genome size is expressed in millions of base pairs (Mbp) per haploid genome. Where a species' name has changed due to the taxonomic revision (Carvalho, F.A. 2013 onward), the name used in the original publication is given in brackets. Four C-values were obtained in August 2013 in the lab of Ray Ming at the University of Illinois in Urbana-Champaign from plants grown from the same seed lots and following the methods of Gschwend et al. (2013); other chromosomes numbers come from Tina Kyndt, Dept. of Molecular Biotechnology, Ghent University (personal communications, August and November 2013). RUG refers to E. H Romeijn-Peeters (Kyndt et al. 2005: Table 1). HCAR refers to the accession number of the Clonal Germplasm Repository for Tropical and Subtropical Crops in Hilo, Hawaii, USA.

Species	Voucher	Chromosome number (size)	Genome size	References
Cylicomorpha parviflora Urb.	Carvalho, 2238 (M)	$2n = 18$		this study
			968 Mbp	R. Ming, from same seed lot
Jarilla caudata (Brandegee) Standl.	Carvalho, 2240 (M)	$2n = 14$		this study
			924 Mbp	R. Ming, from same seed lot
Jarilla heterophylla (Cerv. ex La Llave) Rusby	Carvalho, 2239 (M)	$2n = 14$		this study
			836 Mbp	R. Ming, from same seed lot
Horovitzia cnidoscoloides (Lorence & Torres Colín, R.) V.M.Badillo	Carvalho, 2242 (M)	$2n = 16$		this study
			401 Mbp	R. Ming, from same seed lot
Carica papaya L.	Not cited	$2n = 18$		Heilborn 1921; Simmonds 1954; Joshi & Ranjekar 1982
	Not cited	$2n = 18$ (1-4.23 µm)		Datta, 1971
	Not cited		372 Mbp	Arumuganathan and Earle 1991
	Not cited	$2n = 18$		Costa et al. 2008
	Not cited	$2n = 18$ (1.52-2.29 µm)		Damasceno et al. 2009
	Not cited	$2n = 18$	318 Mbp	Araújo et al. 2010
	HCAR 320		442.5 Mbp	Gschwend et al. 2013
	RUG 57 (GENT)	$2n = 18$		T. Kyndt
Jacaratia spinosa (Aubl.) A.DC. (Carica dodecaphylla Vell.)	Not cited	$2n = 18$		Kumar & Srinivasan 1944
	Not cited	$2n = 18$		Silva et al. 2012
	HCAR 227		513.6 Mbp	Gschwend et al. 2013

88

Vasconcellea candicans A. DC.	RUG113 (GENT)	$2n = 18$		T. Kyndt
Vasconcellea cauliflora (Jacq.) A.DC.	Not cited	$2n = 18$		Caetano et al. 2008
	RUG 284 (GENT)	$2n = 18$		T. Kyndt
Vasconcellea glandulosa A.DC.	HCAR 300		534.9 Mbp	Gschwend et al. 2013
Vasconcellea goudotiana Triana & Planch.	Not cited	$2n = 18$		de Zerpa 1959
	Not cited	$2n = 18$		Costa et al. 2008
	Not cited	$2n = 18$		Caetano et al. 2008
(*Carica goudotiana* (Triana & Planch.) Solms)	Not cited	$2n = 18$		Silva et al. 2012
	HCAR 167		607 Mbp	Gschwend et al. 2013
	RUG285 (GENT)	$2n = 18$		T. Kyndt
Vasconcellea longiflora (V.M.Badillo) V.M.Badillo	Not cited	$2n = 18$		Caetano et al. 2008
	RUG228 (GENT)	$2n = 18$		T. Kyndt
Vasconcellea microcarpa (Jacq.) A.DC. (*Carica microcarpa* Jacq.)	Not cited	$2n = 18$		de Zerpa 1959
Vasoncellea monoica (Desf.) A.DC.	Not cited	$2n = 18$		de Zerpa 1959
	Not cited	$2n = 18$ (1.35-2.49 µm)		Damasceno et al. 2009
(*Carica monoica* Desf.)	HCAR 171		625.9 Mbp	Gschwend et al. 2013
	RUG58 (GENT)	$2n = 18$		T. Kyndt
Vasconcellea pubescens A.DC.	Not cited	$2n = 18$		Heilborn 1921; Costa et al. 2008; Caetano et al. 2008
(*Vasconcellea cundinamarcensis* V.M. Badillo)	Not cited	$2n = 18$ (1.66-2.45 µm)		Damasceno et al. 2009
	HCAR 46		566.7 Mbp	Gschwend et al. 2013
	RUG161 (GENT)	$2n = 18$		T. Kyndt
Vasconcellea palandensis (V.M.Badillo, Van den Eynden & Van Damme) V.M.Badillo	Not cited	$2n = 18$		Caetano et al., 2008
Vasconcellea quercifolia A.St.-Hil.	Not cited	$2n = 18$		Silva et al. 2012
	RS3586	$2n = 18$		Bernadello et al. 1990
	HCAR 226		516.1 Mbp	Gschwend et al. 2013

Vasconcellea *horovitziana* (V.M. Badillo) V.M.Badillo	HCAR 305		557.7 Mbp	Gschwend et al. 2013
Vasconcellea *parviflora* A.DC.	HCAR 180/179		491.5 Mbp	Gschwend et al. 2013
Vasconcellea sphaero-carpa (García-Barr. & Hern.Cam.) V.M.Badillo	Not cited	2*n* = 18		Caetano et al. 2008
Vasconcellea stipulata (V.M.Badillo) V.M.Badillo	HCAR 177 RUG 55(GENT)	2*n* = 18	520.1 Mbp	Gschwend et al. 2013 T. Kyndt
Vasconcellea pulchra (V.M.Badillo) V.M. Badillo	HCAR 267		554.6 Mbp	Gschwend et al. 2013
Vasconcellea weber-baueri (Harms) V.M.Badillo	RUG10 (GENT)	2*n* = 18		T. Kyndt
Vasconcellea × *heilbornii* (V.M. Badillo) V.M.Badillo	Not cited RUG198 (GENT)	2*n* = 18 2*n* = 18		Heilborn 1921 T. Kyndt

Acknowledgements

For assistance in the lab, we thank Dr. Martina Silber and Sinem Demirkaya. This study was financially supported by a doctoral fellowship (CNPq project 290009/2009-0) to Fernanda A. Carvalho and a grant from the Deutsche Forschungsgemeinschaft (DFG RE 603/13-1) to Susanne Renner.

References

Araújo, F.S., Carvalho, C.R. & Clarindo, W.R. (2010) Genome size, base composition and karyotype of *Carica papaya* L. *Nucl.* 53, 25–31

Arumuganathan, K. & Earle, E.D. (1991) Nuclear DNA content of some important plant species. *Plant Mol. Biol. Report.* 9, 415–415

Bernardello, L.M., Stiefkens, L.B. & Piovano, M.A. (1990) Números cromosómicos en dicotiledóneas Argentinas. *Boletín la Soc. Argentina Botánica* 26, 149–157

Caetano, C.M., Burbano, T.C.L., Sierra, C.L.S., Tique, C.A.P. & Nunes, D.G.C. (2008) Citogenética de especies de *Vasconcellea* (Caricaceae). *Acta Agron* 57, 241–245

Carvalho, F.A. & Renner, S.S. (2012) A dated phylogeny of the papaya family (Caricaceae) reveals the crop's closest relatives and the family's biogeographic history. *Mol. Phylogenet. Evol.* 65, 46–53

Carvalho, F.A. (2013) e-Monograph of Caricaceae. Version 1 [Database continuously updated]. Available at: http://herbaria.plants.ox.ac.uk/bol/caricaceae

Costa, F.R., Pereira, T.N.S., Hodnett, G.L., Pereira, M.G. & Stelly, D.M. (2008) Fluorescent in situ hybridization of 18S and 5S rDNA in papaya (*Carica papaya* L.) and wild relatives. *Caryologia* 61, 411–416

Damasceno, C., Pedro, J., Rabelo, F., Santana, T.N., Neto, M.F. & Pereira, M.G. (2009) Karyotype determination in three Caricaceae species emphasizing the cultivated form (*C. papaya* L.). *Caryologia* 62, 10–15

Datta, P.C. (1971) Chromosomal biotypes of *Carica papaya* Linn. *Cytologia* 36, 555–562

deZerpa, D.M. (1959) Citologia de hibridos interespecificos en *Carica*. *Agron. Trop.* 3, 135–144

Gschwend, A.R., Wai, C.M., Zee, F., Arumuganathan, A.K. & Ming, R. (2013) Genome size variation among sex types in dioecious and trioecious Caricaceae species. *Euphytica* 189, 461–469

Heilborn, O. (1921) Taxonomical and cytological studies of cultivated Ecuadorian species of *Carica*. *Arkiv för Botanik* 17(12), 1–16

Joshi, C.P. & Ranjekar, P.K. (1982) Visualization and distribution of heterochromatin in interphase nuclei of several plant species as revealed by a new Giemsa Banding technique. *Cytologia* 47, 471–480

Kumar, L.S.S. & Srinivasan, V.K. (1944) Chromosome number of *Carica dodecaphylla* Vell. *Curr. Sci.* 13, 15

Luo, M.C., Deal, K.R., Akhunov, E.D., Akhunova, A.R., Anderson, O.D., Anderson, J.A., Blake, N., Clegg, M.T., Coleman-Derr, D., Conley, E.J., Crossman, C.C., Dubcovsky, J., Gill, B.S., Gu, Y.Q., Hadam, J., Heo, H.Y., Huo, N., Lazo, G., Ma, Y., Matthews, D.E., McGuire, P.E., Morrell, P.L., Qualset, C.O., Renfro, J., Tabanao, D., Talbert, L.E., Tian, C., Toleno, D.M., Warburton, M.L., You, F.M., Zhang, W. & Dvorak, J. (2009) Genome comparisons reveal a dominant mechanism of chromosome number reduction in grasses and accelerated genome evolution in Triticeae. *Proc. Natl. Acad. Sci. USA* 106, 15780–15785

Lysak, M.A, Berr, A., Pecinka, A., Schmidt, R., McBreen, K. & Schubert, I. (2006) Mechanisms of chromosome number reduction in *Arabidopsis thaliana* and related Brassicaceae species. *Proc. Natl. Acad. Sci. USA* 103, 5224–5229

Mandáková, T. & Lysak, M.A. (2008) Chromosomal phylogeny and karyotype evolution in x = 7 crucifer species (Brassicaceae). *Plant Cell* 20, 2559–2570

Silva, E.N. da, Neto, M.F., Pereira, T.N.S. & Pereira, M.G. (2012) Meiotic behavior of wild Caricaceae species potentially suitable for papaya improvement. *Crop Breed. Appl. Biotechnol.* 12, 52–59

Silva, N., Mendes-Bonato, A.B., Sales, J.G.C. & Pagliarini, M.S. (2011) Meiotic behavior and pollen viability in *Moringa oleifera* (Moringaceae) cultivated in southern Brazil. *Genet. Mol. Res.* 10, 1728–1732

Simmonds, N.W. (1954) Chromosome behavior in some tropical plants. *Heredity* 8, 139–146

Yang, L., Koo, D.H., Li, D., Zhang, T., Jiang, J., Luan, F., Renner, S.S., Hénaff, E., Sanseverino, W., Garcia-Mas, J., Casacuberta, J., Senalik, D. A, Simon, P.W., Chen, J. & Weng, Y. (2013) Next-generation sequencing, FISH mapping, and synteny-based modeling reveal mechanisms of decreasing dysploidy in *Cucumis*. *Plant J.* 1–15

Yogeeswaran, K., Frary, A., York, T.L., Amenta, A., Lesser, A.H., Nasrallah, J.B., Tanksley, S.D. & Nasrallah, M.E. (2005) Comparative genome analyses of *Arabidopsis* spp.: Inferring chromosomal rearrangement events in the evolutionary history of *A. thaliana*. *Genome Res.* 15, 505–515

VI. General Discussion

Taxonomy in the Electronic Age and an e-Monograph of Caricaceae

Bioinformaticians have developed many tools for dealing with the kinds of data needed for taxonomic work on higher plants. Information Technology (IT)-infrastructure or Cyber-infrastructure is a combination of databases, network protocols, and computational services that bring together information and computational tools to perform research in an information-driven world (Stein 2008). This new infrastructure comprises at least four components, (*i*) data infrastructure for storing, integrating and retrieving essential information; (*ii*) computational infrastructure for manipulating and analyzing data sets; (*iii*) communication infrastructure for interconnecting the computational and data resources; and (*iv*) human infrastructure for using the resources available (data and bio-informatic tools), thus, facilitating collaboration among researchers (Stein 2008). In the context of taxonomic research, the IT-infrastructure ideally provides a set of tools not only for documenting and disseminating knowledge on species diversity, but also for generating new knowledge about species.

During the last two years, I digitized specimens deposited in 20 herbaria and gathered taxonomic data available on the Caricaceae family, including web links to the supportive literatures and high-resolution images of type specimens, where freely available. This process resulted in a database with c. 10,000 images and information on 4,419 specimens, representing 2,598 collections. I used Google Earth and other online resources to georeference as many as possible of the collections with locality information (c. 2,000). I also compiled information on all available names, and resolved nomen-clatural issues. For example, I established the use of *Vasconcellea pubescens* A.DC. (accepted in Flora Mesoamericana), resolved a confusion among the correct names of two species of *Jarilla* (Chapter III). Others will be formally published together with the complete monograph (examples given in the Appendix) among others in the open-access journal *PhytoKeys*, a journal at the forefront of new technologies that better integrate taxonomic data with aggregators of information, such as the Plant List, Global Biodiversity Information Facility, and Encyclopedia of Life (Penev et al. 2010).

In order to delimit species boundaries, I initially separated the specimens according to the presence of pistillate (female) or staminate (male) flowers, or fruits. I relied only on fertile material because the presence of reproductive structures made specimens reliably comparable, although of course I also considered vegetative characters, especially leaf/leaflet shape and venation, which demonstrated to be very useful to distinguish some species. Next, I grouped specimens by geography and then separated clusters with obvious differences. Similarities in distribution, habitat, and morphology were then used to finalize piles of male and female specimens representing presumed biological species (gene pools). I inferred the habitat occupied by each pile of specimens by overlapping the distribution maps and GIS datasets (climate, pH of top soil, land cover); information about accompanying plants or vegetation types on labels were also considered.

Badillo (1993) dealt with 155 names, and my database contains 233 names, a difference of 78 names in 10 years. This difference is due to 16 new names published after 1993 (Badillo 2000; Badillo et al. 2000) and 62 names that I found in online databases or in old literature and that Badillo overlooked because he had no access to these sources of data. This is a result of mass digitization programs of biological collections and books. The better access to the old literature and collections data, together with taxon search tools implemented in digital libraries, such as the Biodiversity Heritage Library (http://www.biodiversitylibrary.org/advsearch) are helping taxonomists to unearth names never indexed or cited before. On the one hand, the massive digitization programs are very positive because they are speeding up the process of acquiring biodiversity data and make taxonomy more democratic because also workers at institutions without rich old libraries have access to rare or ancient texts. On the other hand, these programs are increasing the problem of synonyms because more and more invalidly published names, including spelling permuations and *nomina nuda* are being found and disseminated. For example, two specimens digitized for the Latin America Plants Initiative are stored in JSTOR Plant Science, as *Jacaratia harleyi* Badillo, a name never published, although Badillo used it on some labels. This and other problems (see below) reflect the generally poor (electronic) "communication" among the databases of different institutions. The problem with the name *J. harleyi* is now fixed in the Kew database, where the specimen is deposited and where I visited in February 2013. However, the error persists in JStor Plants (http://plants.jstor.org/specimen/k000500514, accessed Dec. 12, 2013).

A major problem in building the Caricaceae's database was to gather data from different online resources, which continue to use different standards and field definitions. As feared by Stein (2008, p. 686), "the most likely map of the biological cyberinfrastructure that is coming in the immediate future is an archipelago of islands; each discipline using a grid that is internally consistent, but effectively isolated from the others; " nevertheless proposals to standardize biodiversity databases already exist (e.g., HISPID; Conn 1995 and Darwin Core; Wieczorek et al. 2012). New software and platforms are being developed each year by different institutions, but the communication among them is not being improved at the same pace. Database mapping can be used to integrate two distinct data models. However, if the same piece of information is digitized slightly different among institutes, queries that address multiple databases may not be able to be adequately solved (Willemse et al. 2008). Standardization in data entry would increase the value of freely available biodiversity data by facilitating the use and re-use, distribution, and integration of this information.

As I have discussed in Chapter II of this thesis, monography is the only way to produce a well-resolved, expert-vetted nomenclature for an entire group with information on the distribution and morphology of all species and a complete alphabetical list of accepted and synonymized names. However, high-quality data produced by taxonomists in revisions and monographs are of little use unless widely accessible and easily understandable. This is especially important for economically important groups, which are often also groups with a high rate of nomenclatural changes (as is the case for Caricaceae). Open-access information from my e-monograph that includes organized set of data and images for the Caricaceae benefits the scientific community broadly as well as those working on food or medicinal properties of papaya and its relatives. This includes researchers focusing on papaya genomics, ecologists, and breeders (e.g., Scheldeman et al. 2007; Gschwend et al. 2013; Coppens d'Eeckenbrugge et al. 2014).

Phylogenetic Relationships Within Caricaceae

The molecular phylogeny of Caricaceae I produced (Chapter IV) is the first to include all species from all six genera sampled for both nuclear and plastid sequences. The phylogeny shows, with high support, that all genera with more than one species are monophyletic. The deepest divergence in the family is between the African *Cylicomorpha* and the Neotropical genera. In the Neotropics, *Jacaratia* and *Vasconcellea* form a clade sister to all other four genera, which includes the monotypic *Carica papaya*. Within *Jacaratia* and *Vasconcellea*, species relationships are not well resolved although there are a few well-supported groups. *Vasconcellea microcarpa* is a highly variable species distributed from the mouth of the Amazonas River to the other side of the Andes, along the Pacific coast. There are four well-defined morphotypes formally named as *Vasconcellea microcarpa* (Jacq.) A.DC. ssp. *microcarpa*, *V. microcarpa* ssp. *baccata* (Heilborn) Badillo (Western Andes), *V. microcarpa* ssp. *pilosa* Badillo (Northern Andes) and *Carica microcarpa* ssp. *australis* Badillo (Central Bolivia), which are indistinguishable along the contact zone (i.e., the eastern bottom of the Andes). Further sampling of *Vasconcellea microcarpa* at the population level will be necessary to better understand this species and would also contribute to our understanding of the historical connections of Trans-Andean clades.

The closest relatives of papaya are four species distributed from Mexico to El Salvador, with the unilocular ovary being an apomorphy of the group (all remaining Caricaceae have 5-locular ovaries). Within the sister group of papaya, the three species of *Jarilla* are sister of *Horovitzia cnidoscoloides*, which implies that the latter could be included in *Jarilla* to make the classification more informative in terms of relationships.

However, I decided to retain the single species of *Horovitzia* as a separate genus due to its remarkable morphology: stinging hairs, subcapitate stigma, and all anthers with two thecae. The three *Jarilla* species are glabrous or slightly pubescent (trichomes simple), have long stigmas, and the superior anther with only one theca. In addition, *Horovitzia* occupies a very special ecological niche not shared with any other Central American Caricaceae. It is endemic to cloud forests of Sierra de Juaréz in Oaxaca, Southern Mexico from 1000–1500 m high.

The fact that the closest relatives of papaya are all endemic to Central America has implications for the place of origin and domestication of papaya, which has been deduced either from centers of early human civilizations or from centers of Caricaceae species diversity. Classical studies on crops domestication proposed Mexico as the place of origin (De Candolle 1883; Solms-Laubach 1889; Vavilov 1987), but others suggested northwestern South America, because this area is the center of diversity of Caricaceae, with most of *Vasconcellea* species endemic to the northern Andes region (Badillo 1971; Prance 1984). According to Vavilov (1987) there are two centers of crop domestication in the Neotropics, and papaya could have been domesticated either in the Peruvian/Bolivian center by the Incas or in the (southern) Mexican center by the Mayas. Based on my findings, only the latter now remains plausible.

The wild form of papaya, which has much smaller and rounded fruits (maximum 7 cm in diameter when mature) and also much thinner mesocarp (less the 1 cm) than the cultivated form, was first described based on a collection from El Realejo, today in Chinandega, northwestern Nicaragua (Hooker and Arnott 1840). Since then, other populations of the small papayas have been found in open areas and forests edges in the lowlands, from southern Mexico to northern Costa Rica. Fruits are tasty, but less sweet than the cultivated form. Since wild papayas have never been found outside Central America, and since the closest relatives of *C. papaya* occur only in Mexico, Guatemala and El Salvador, domestication by Mayan Indians in the Mesoamerican lowlands appears likely, although there is no direct archaeological evidence.

Historical Biogeography and the Importance of Habitat Diversity in the Evolution of Caricaceae

My ancestral area reconstruction, combined with the molecular clock-dating analysis, indicates that a transoceanic dispersal occurred from Africa to Central America during the Late Eocene, around 35 Mya. This dispersal event could have involved a floating island carried from the Congo delta by the North Atlantic Equatorial current as also suggested for other plants and animals (e.g., Houle 1999; Renner 2004; Vidal et al. 2008). Caricaceae have soft, fleshy fruits not suitable for water dispersal, but seeds could have been carried on such floating vegetation. Even if transport took several weeks, seeds might not have germinated because germination in the family is slow and erratic, which has been attributed to inhibitors present in the sarcotesta (Tokuhisa et al. 2007).

From Central America, Caricaceae then reached South America sometime between 27 and 19 Mya, probably by island hopping across the Panamanian Isthmus. Traditional as well as the most recent palaeo-geological studies (Coates et al. 2004; Farris et al. 2011; Montes et al. 2012) demonstrate increasingly shorter distances between Central and South America during the late Oligocene to early Miocene, which facilitated range expansion from Central America to Northern South America where Caricaceae established and started to diversify gradually expanding its distribution to the south. Such Oligocene/Miocene island hopping between Mexico and Colombia probably also explains the ranges of other plant groups distributed along South and Central America such as *Hechtia* (Bromeliaceae; Givnish et al. 2011) and *Copernica* (Arecaceae; Bacon et al. 2013).

The African Clade (*Cylicomorpha*)

Although the split between the African and the Neotropical Caricaceae is quite old, my molecular clock dating indicates that the history of the two extant African species is young: *Cylicomorpha solmsii* and *C. parviflora* diverged from each other during the Plio-Pleistocene boundary, around 2.8 (0.6–5.2) Mya (Chapter II). At this time, Africa was characterized by extreme climate variability with alternating periods of high moisture levels and extreme aridity (Sepulchre et al. 2006; Trauth et al. 2009) with a change from wet to dry conditions occurring during the Late Pliocene, between 4 and 3 Mya (Sepulchre et al. 2006). Today, both species are big trees reaching up to 40 m tall and occur in rainforest paths in West Africa (*C. parviflora*, 500–2000 m high) and East Africa (*C. solmsii*, 400–1200 m high; Fig. 1). Their modern ranges clearly result from the fragmentation of evergreen tropical forests during the Pliocene, and their divergence time matches the inferred ages of other East and West African rainforest clades (Bowie et al. 2004; Couvreur et al. 2008, 2011).

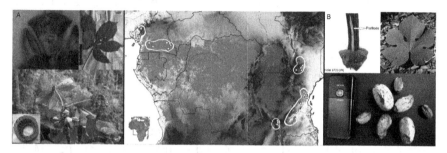

Fig. 1. The distribution of *Cylicomorpha solmsii* in West Africa and *C. parviflora* in the East Africa (*yellow outlines*), and the current distribution of evergreen forests in Central Africa based on a global land cover classification derived from images of AVHHR satellites acquired between 1981 and 1994 (http://glcf.umd.edu/data/landcover/). (**A**) *Cylicomorpha solmsii*: calyx and a deeply lobed leaf (Preuss, 489 [K]), a piece of the hollowed stem (A.J.M. Leeuwenberg, 9745 [P]), and a living branch with fruits. (**B**) *Cylicomorpha parviflora*: calyx and pistillode (C. Holst 8723 [W]), a living leaf, and fruits. The calyx margins of both species are slightly undulate or entire, an apomorphy of the genus. The hollowed trunk is a trait found only in the two species of *Cylicomorpha* and in *Carica papaya*. Photos of living material of *C. solmsii* by Jean Paul Ghogue and of *C. parviflora* by Mark Nicholson.

The Central American Clade

The main features of Sierra Madre mountain range in Mexico and the volcanic belt that extents from Mexico along the width of Central America, developed from the Oligocene until the Pliocene, and triggered the establishment of modern biomes in West and Central Mexico (Becerra 2005; Ferrari et al. 2012). The diversification of the papaya clade began in the late Oligocene around 25 Mya when *Carica papaya* diverged from its closest relatives, followed by the split between

Horovitzia cnidoscoloides and *Jarilla* around 18 Mya (Chapter II). These two splits occurred after the raising of the Sierra Madre Occidental and the western portion of the Trans-Mexican Volcanic Belt (TMVB) during the late Oligocene and early Miocene (Gómez-Tuena et al. 2007; Ferrari et al. 2012). The divergence between the three species of *Jarilla* and *Horovitzia* is estimated to have occurred between 7 and 3 Mya, during a major episode of intense volcanism along the TMVB that established the west-eastern highland corridor in Central Mexico (c. 7.5–3 Mya; Gómez-Tuena et al. 2007). Today, the Sierra Madre Occidental and the TMVB block the cold fronts from the North and shelter the seasonally dry forests, one of the most extensive types of vegetation in Mexico (Becerra 2005). The establishment of these forests along the Pacific coast may have influenced the differentiation of *Jarilla chocola*, which is widespread in this biome from northern Mexico to El Salvador, reaching altitudes of maximum 1000 m. Its distribution along the pacific coast is delimited by the Sierra Madre Occidental (Fig. 2). Two ancient isolated populations of *Jarilla caudata* and *J. heterophylla* may have come in contact after the establishment of the east-west highland corridor formed by the TMVB, which would explain the co-occurrence of these two species in dry habitats at higher altitudes (1500–2000 m; Fig. 2). Other taxa co-occurring in the same region also diverged during the Miocene-Pliocene boundary, corroborating the importance of the processes of mountain building, specially the formation of the TMVB in the diversification of organisms from central Mexico (*Bursera*: Becerra 2005; *Aphelocoma* jays: McCormack et al. 2011; alligator lizards: Bryson Jr & Riddle 2012).

The Mostly South American Clade *Vasconcellea/Jacaratia*

The mostly South American genera, *Vasconcellea* (20 species) and *Jacaratia* (seven species), shared a common ancestor during the Miocene (c. 19 Mya) when global climatic changes were causing expansion of savannas and when the Andes were formed (Hoorn et al. 2010; Pound et al. 2011). Diversification of *Vasconcellea*, the largest genus of Caricaceae, begun around 14 Mya coinciding exactly with the first peak of northern Andean uplift during the late Miocene (Hoorn et al. 2010).

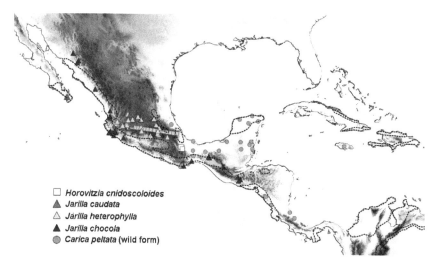

□ *Horovitzia cnidoscoloides*
▲ *Jarilla caudata*
△ *Jarilla heterophylla*
▲ *Jarilla chocola*
● *Carica peltata* (wild form)

Fig. 2. Distribution map of the five Central American species. *Horovitzia cnidoscoloides* occurs in the cloud forests of Sierra de Juarez characterized by high precipitation, annual mean temperature varying from 19°C to 23°C and no seasonality. *Jarilla chocola* occurs in the seasonally dry forests (*dotted outline; according to Pennington et al. 2000*) along the Pacific, characterized by relatively low precipitation, high seasonality, and annual mean temperature varying from 20°C to 28°C. The wild form of *Carica papaya* occurs in the lowlands in relatively dry areas with annual mean temperature varying from 24°C to 29°C. *Jarilla caudata* and *J. heterophylla* overlap along the Trans-Mexican volcanic Belt (*dotted green line*) in similar environmental conditions (i.e., relatively low precipitation, high seasonality, and annual mean temperature varying from 16°C to 22°C).

In the Andes, *Vasconcellea* occupies a broad range of habitats, including evergreen forests, cloud forests and semi-desert areas, suggesting a strong influence of the environment in the diversification of the group, that is in species radiating into different habitats (Fig. 3). However, in other well-supported sister species pairs, such as *V. pulchra* and *V. longiflora*; *V. sphaerocarpa* and *V. goudotiana*; and *V. pubescens* and *V. goudotiana*, each member of a pair occupies similar environmental conditions, showing that niche conservatism also plays a role.

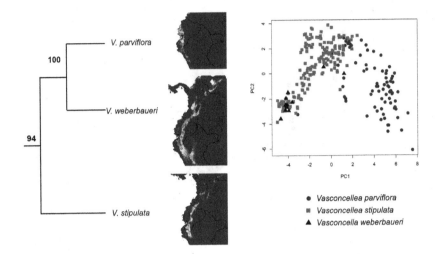

Fig. 3. Species divergence in *Vasconcellea* seems to have been driven by both *climatic niche conservatism* as a consequence of allopatric speciation and by *climatic niche divergence* as a consequence of new habitats occupied after the Andean uplift. The figure on the left summarizes climate niches occupied by three species that constitute one a well-supported clade. The maps are climatic niche models estimated with the maximum entropy algorithm MaxEnt (Phillips and Dudík 2008) using non-correlated climatic variables from worldclim (30 sec resolution; http://www.worldclim.org/). On the right, a Principal Component Analysis (PCA) was performed on a correlation matrix of all 20 Bioclim variables. The two first axes (PC1 and PC2) account for c. 74% of environmental variation among species. Two well-defined climatic niches are formed, indicating that *V. stipulata* (wet montane forests, sometimes associate to rocks and deep slopes, 1500–2600 m alt.) and *V. weberbaueri* (shady places in wet montane forests, 1500–3000 m alt.) occupy similar conditions, while *V. parviflora* occurs in a different habitat (deciduous or semideciduous forests at 50–1250 m alt.).

Diversification of *Jacaratia* also began during the Mid-Miocene, around 13 Mya, when a warmer and drier climate promoted the expansion of savannas worldwide (Pound et al. 2011). Today *Jacaratia* consists of seven species, six occurring in the lowlands of South and Central America, and one, *J. chocoensis*, occurring at higher altitudes (up to 1500 m alt. in the Andes). The two Central American species, *J. dolichaula* and *J. mexicana*, are embedded in a South American clade, suggesting that they reached Central America from South America (Chapter II). They occur in distinct habitats (Fig. 4), indicating again that niche divergence is related to speciation in Central America: *J. mexicana* is widely distributed in seasonally dry forests, while *J. dolicaula* occurs exclusively in lowland rainforests of Central America. In South America, the dry climate cycles may also have favored the occupation of dry habitats by *J. corumbensis*. This species presents a disjunct distribution with two main populations, one in Northeast Brazil in Caatinga forest and the other one in midwestern South America in Chaco forest (Fig. 4). Caatinga is a seasonally tropical dry forest characterized by low temperature seasonality, but with an extreme dry season; Chaco is a subtropical extension of a temperate formation in midwest South America and is characterized by regular frosts (high temperature seasonality), and long dry season (Pennington et al. 2000). My dating analyses, including one accession of *J. corumbensis* from Caatinga and one from the Midwest South America, indicates that the two populations diverged from each other during the Pleistocene, c. 0.8 Mya (Fig. 4)., a period known for its rapid changing climatic cycles and related to events of expansion and retraction of dry forests in South America (Pennington et al. 2000). Further phylogeographic studies of *Jacaratia* species could provide further insights on the historical biogeography of the dry forest regions of South America. In addition to the historical connections between Chaco and Caatinga, the connections between Atlantic forest and Amazonia could also be better understood through population studies of *J. heptaphylla* (Atlantic forest), and *J. spinosa* (Amazonia and Atlantic forest).

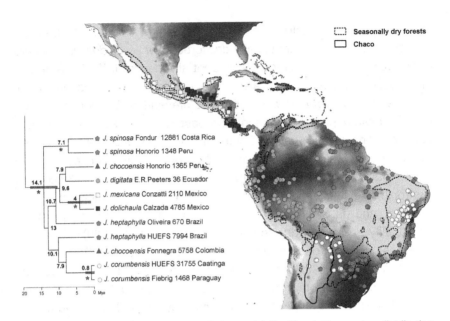

Fig. 4. Chronogram and distribution of *Jacaratia*. On the right, species distribution plotted on a layer of precipitation seasonality with cold colors representing regions with low seasonality (Bio 15 at 2.5 min resolution; http://www.worldclim.org/). The outlines represent the seasonally dry forests (*dotted outlines*) and the Chaco forest (*solid outline*) based on Pennington et al. (2000; with permission). *Jacaratia mexicana* (from Mexico to Nicaragua) and *J .corumbensis* from the Caatinga (Ne Brazil) are associated with seasonally dry forests, with relatively high precipitation seasonality. *Jacaratia dolichaula* seems to replace *J. mexicana* in the evergreen forests from southern Mexico to Panama. *Jacaratia digitata* occurs in the western Amazon in regions with low (or no) precipitation seasonality. *Jacaratia chocoensis* is rarely collected, and its distribution and environmental requirements are therefore unclear. *Jacaratia spinosa* has a larger range and occurs in evergreen forests with low to medium precipitation seasonality. *Jacaratia heptaphylla* occurs along the Atlantic coast in southeast Brazil, and its current narrow distribution range may be due to degradation of the Atlantic forest. Chronogram for *Jacaratia* under a strict clock model applied to two plastid genes (*mat*K and *rbc*L). On the left, an updated chronogram generated by adding new sequences to the same dataset of Carvalho & Renner (2012; Chapter IV). Values on nodes are the Bayesian divergence times (in million years). * indicates nodes with posterior probabilities higher than 0.94.

Evolution of Chromosome Numbers in the Caricaceae

Many factors influence changes in chromosome size, shape, and numbers (which together are make up a species' karyotype). Chromosomes can be altered by insertions of parts of other chromosomes or entire other chromosomes as well as by rearranging parts within or between chromosomes (Schubert and Lysak 2011). The new chromosome counts of the closest relatives of papaya, which were unknown before my work and which were first brought into cultivation at the Munich Botanical Garden as a result of my fieldwork, indicate a reduction of chromosomes number in the most recent common ancestor of *Horovitzia* ($2n = 16$) and *Jarilla* ($2n = 14$), while all remaining Caricaceae studied so far, including papaya, have $2n = 18$ (Chapter V). The different chromosomes lengths seen in the two species of *Jarilla* suggest that its large chromosomes may be the result of chromosomal rearrangements, perhaps fusions. Lysak et al. (2006), using chromosome painting, showed successive events of inversions, translocations, and fusion which led to a reduction in the chromosome number from $n = 8$ to $n = 5$ within the genus *Arabidopsis*, another member of the Brassicales. Further studies using chromosome painting techniques should improve our understanding on the karyotype evolution of the papaya clade. My results have important implication for plant breeders already now since it will be not possible to transfer useful traits, such as cold tolerance, from its wild relatives to papaya via natural crossings: *Carica papaya* has $n = 9$ and its closest relatives seem to have $n = 8$ (Chapter V).

General Conclusions

The electronic monograph that I produced is one of the first (if not *the* first) that have been completed. It is an example of how new technologies can be used to speed up taxonomy and most importantly monography, thus reducing the problem of synonymous names and increasing the quality of data in other aggregators of information, such as EOL, GBIF, TROPICOS, and BHL. My e-monograph is also an example of how the results of taxonomic work can be disseminated efficiently and – eventually – in an interactive manner although this is not yet implemented as I am the sole responsible author of this doctoral research. Besides being an example of e-monography, the economic importance of the Caricaceae itself justifies having all information freely available. The data will benefit a large constituency, including herbaria curators, researchers focusing on papaya genomics, the papain industry, ecologists, breeders, and the nonscientific public. Today, a big challenge for taxonomists is to make better use of all the information freely accessible on the web to improve and generate new taxonomic information, and to spread the products of systematic research to non-systematists. Electronic monographs will greatly improve access to the knowledge about species, while at the same time feeding other databases with invaluable information for scientific research, society, and industry.

The complete species-level phylogeny of the Caricaceae resolved the evolutionary relationships among the family's genera, although species relationships within the largest genera (*Vasconcellea* and *Jacaratia*) remained unclear. The two African species (*Cylicomorpha solmsii* and *C. parviflora*) are sister to Neotropical genera, and *Vasconcellea* and *Jacaratia* (mostly from South America) form a clade sister to a Central American group, which includes papaya. The closest relatives of papaya are three perennial herbs (*Jarilla chocola, J. heterophylla,* and *J. nana*) and one small tree with a thin and soft trunk, *Horovitzia cnidoscoloides*, all restricted to the region from northern Mexico to El Salvador. This region coincides with the distribution of the wild form of *Carica papaya* and thus is the most plausible place of origin of this important crop, probably domesticated by one of the Mesoamericans civilizations (Mayas or Olmecs). As a general contribution, my study shows the importance of evaluating crops in a phylogenetic perspective. Besides inferences about the origin of crops without archeological records (such as papaya), there are also implications for plant breeders. During decades, plant breeders have tried in vain to cross papaya with species of a phylogenetically distant group (*Vasconcellea*), while from now on their efforts might include genetic comparisons with *Jarilla* and *Horovitzia*.

The biogeographic history of Caricaceae involves long-distance dispersal from Africa to Central America c. 35 Mya and range expansion through island hopping from Mexico to Colombia prior to the full closure of the Panamanian land bridge. In Africa, the two species of *Cylicomorpha* diverged from each other during the Plio-Pleistocene boundary, around 3 Mya, when the African savannas expanded due to dry climatic conditions. Diversification of *Vasconcellea*, the largest genus of the family, is related to the peak of the northern Andean orogeny (c. 14 Mya), while diversification of *Jacaratia* appears to be linked to the expansion of drought-adapted vegetation during the Late Miocene (c. 12 Mya). However, more detailed studies at the population level are needed to solve species relationships within these genera and to make better inferences on external factors that may have shaped the evolution of these two lineages. Polyploidy is not a mechanism that has caused speciation in Caricaceae, and chromosomal rearrangements (fusion) seem to have played a role especially in the papaya clade. Finally, the niche modeling carried out as part of my research shows that ecological divergence in habitat types but also classic allopatric speciation were both important factors driving the diversification of Caricaceae.

References

Aguirre, A., Vallejo-Marín, M., Salazar-Goroztieta, L., Arias, D.M. & Dirzo, R. (2007) Variation in sexual expression in *Jacaratia mexicana* (Caricaceae) in Southern Mexico: Frequency and relative seed performance of fruit-producing males. *Biotropica* 39, 79–86

Anducho-Reyes, M.A., Cognato, A.I., Hayes, J.L. & Zúñiga, G. (2008) Phylogeography of the bark beetle *Dendroctonus mexicanus* Hopkins (Coleoptera: Curculionidae: Scolytinae). *Mol. Phylogenet. Evol.* 49, 930–940

Antoine, P.O., Marivaux, L., Croft, D.A., Billet, G., Ganerød, M., Jaramillo, C., Martin, T., Orliac, M.J., Tejada, J., Altamirano, A.J., Duranthon, F., Fanjat, G., Rousse, S. & Gismondi, R.S. (2012) Middle Eocene rodents from Peruvian Amazonia reveal the pattern and timing of caviomorph origins and biogeography. *Proc. R. Soc. B* 279 1319–1326

Antonelli, A., Nylander, J.A.A., Persson, C. & Sanmartín, I. (2009) Tracing the impact of the Andean uplift on Neotropical plant evolution. *Proc. Natl. Acad. Sci. USA* 106, 9749–9754

Aradhya, M.K., Manshardt, R.M., Zee, F. & Morden, C.W. (1999) A phylogenetic analysis of the genus *Carica* L. (Caricaceae) based on restriction fragment length variation in a cpDNA intergenic spacer region. *Genet. Resour. Crop Evol.* 46, 579–586

Bacon, C.D., Mora, A., Wagner, W.L. & Jaramillo, C.A. (2013) Testing geological models of evolution of the Isthmus of Panama in a phylogenetic framework. *Bot. J. Linn. Soc.* 171, 287–300

Badillo, V.M. (1971) Monografia de la familia Caricaceae. Asociación de profesores, Universidad Central de Venezuela, Maracay. 220 pp.

Badillo, V.M. (1993) Caricaceae. Segundo Esquema. *Rev. la Faculdad Agron. la Univ. Cent. Venez.* 43, 1–111

Badillo, V.M. (2000) *Vasconcella* St.-Hil. (Caricaceae) con la rehabilitacion de este ultimo. *Ernstia* 10, 74–79

Badillo, V.M., Van den Eynden, V. & Van Damme, P. (2000) *Carica palandensis* (Caricaceae), a new species from Ecuador. *Novon* 10, 4

Baker, H.G. (1976) "Mistake" pollination as a reproductive system with reference to the Caricaceae. In: Burley, J., Styles, B.T. (eds) Tropical Trees. Variation, Breeding and Conservation. Academic Press, London, pp. 161–169

Bawa, K.S. (1980) Mimicry of male by female flowers and intrasexual competition for pollintors in *Jacaratia dolichaula* (D. Smith) Woodson (Caricaceae). *Evolution* (N.Y.) 34, 467–474

Bebber, D.P., Carine, M.A., Wood, J.R.I., Wortley, A.H., Harris, D.J., Prance, G.T., Davidse, G., Paige, J., Pennington, T.D., Robson, N.K.B. &

Scotland, R.W. (2010) Herbaria are a major frontier for species discovery. *Proc. Natl. Acad. Sci. USA* 107, 22169–71

Becerra, J.X. (2005) Timing the origin and expansion of the Mexican tropical dry forest. *Proc. Natl. Acad. Sci. USA* 102 10919–23

Beilstein, M.A., Nagalingum, N.S., Clements, M.D., Manchester, S.R. & Mathews, S. (2010) Dated molecular phylogenies indicate a Miocene origin for *Arabidopsis thaliana*. *Proc. Natl. Acad. Sci. USA* 107, 18724–8

Belhumeur, P.N., Chen, D., Feiner, S., Jacobs, D.W., Kress, W.J., Ling, H., Lopez, I., Ramamoorthi, R., White, S. & Zhang, L. (2008) Searching the World's Herbaria: A system for visual identification of plant species. In: Forsyth, D., Torr, P. & Zisserman, A. (eds) *ECCV 2008*, Part IV. vol. 5305. Springer, Heidelberg, pp 116–129

Bernardello, L.M., Stiefkens, L.B. & Piovano, M.A. (1990) Números cromosómicos en dicotiledóneas Argentinas. *Boletín la Soc. Argentina Botánica* 26, 149–157

Blagoderov, V., Brake, I., Georgiev, T., Penev, L., Roberts, D., Ryrcroft, S., Scott, B., Agosti, D., Catapano, T. & Smith, V.S. (2010) Streamlining taxonomic publication: a working example with Scratchpads and ZooKeys. *ZooKeys* 28, 17–28

Bowie, R.C.K., Fjeldså, J., Hackett, S.J. & Crowe, T.M. (2004) Molecular evolution in space and through time: mtDNA phylogeography of the Olive Sunbird (*Nectarinia olivacea/obscura*) throughout continental Africa. *Mol. Phylogenet. Evol.* 33, 56–74

Bryson Jr, W.R. & Riddle, B.R. (2012) Tracing the origins of widespread highland species: a case of Neogene diversification across the Mexican sierras in an endemic lizard. *Biol. J. Linn. Soc.* 105, 382–394

Caetano, C.M., Burbano, T.C.L., Sierra, C.L.S., Tique, C.A.P. & Nunes, D.G.C. (2008) Citogenética de especies de *Vasconcellea* (Caricaceae). *Acta Agron* 57, 241–245

Carranza, S., Arnold, E.N., Mateo, J.A. & López-Jurado, L.F. (2000) Long-distance colonization and radiation in gekkonid lizards, *Tarentola* (Reptilia: Gekkonidae), revealed by mitochondrial DNA sequences. *Proc. R. Soc. Lond. B* 267 637–49

Chacón, J., de Assis, M.C., Meerow, A.W. & Renner, S.S. (2012) From East Gondwana to Central America: historical biogeography of the Alstroemeriaceae. *J. Biogeogr.* 39, 1806–1818

Chapman, A.D. (2009) Numbers of Living Species in Australia and the World. 2nd edn. Australian Biological Resources Study (ABRS), Canberra

Chatrou, L.W., Couvreur, T.L.P. & Richardson, J.E. (2009) Spatio-temporal dynamism of hotspots enhances plant diversity. *J. Biogeogr.* 36, 1628–1629

Coates, A.G., Collins, L.S., Aubry, M.P. & Berggren, W.A. (2004) The Geology of the Darien, Panama, and the late Miocene-Pliocene collision of the Panama arc with northwestern South America. *Geol. Soc. Am. Bull.* 116, 1327

Conn, B.J. (ed.) (1995) HISPID – Herbarium Information Standards and Protocols for Interchange of Data. Version 3. Available at http://plantnet.rbgsyd.nsw.gov.au/HISCOM/HISPID/HISPID3/H3.html Accessed 14 December 2013

Coppens d'Eeckenbrugge, G., Drew, R., Kyndt, T. & Scheldeman, X. (2014) *Vasconcellea* for papaya improvement. In: Ming, R., Moore, P.H. (eds) Genet. Genomics Papaya. Springer, New York, pp. 47–79

Coppens d'Eeckenbrugge, G., Restrepo, M.T. & Jiménez, D. (2007) Morphological and isozyme characterization of common papaya in Costa Rica. *Acta Hortic.* 740, 109–120

Costa, F.R., Pereira, T.N.S., Hodnett, G.L., Pereira, M.G. & Stelly, D.M. (2008) Fluorescent in situ hybridization of 18S and 5S rDNA in papaya (*Carica papaya* L.) and wild relatives. *Caryologia* 61, 411–416

Couvreur, T.L.P., Porter-Morgan, H., Wieringa, J.J. & Chatrou, L.W. (2011) Little ecological divergence associated with speciation in two African rain forest tree genera. *BMC Evol. Biol.* 11, 296

Couvreur, T.L.P., Chatrou, L.W., Sosef, M.S.M. & Richardson, J.E. (2008) Molecular phylogenetics reveal multiple tertiary vicariance origins of the African rain forest trees. *BMC Biol.* 6, 54

Damasceno, C., Pedro, J., Rabelo, F., Santana, T.N., Neto, M.F. & Pereira, M.G. (2009) Karyotype determination in three Caricaceae species emphasizing the cultivated form (*C. papaya* L.). *Caryologia* 62, 10–15

De Candolle, A. (1864) Papayaceae. *Prodromus Syst. Nat. Regni Veg.* 15, 413–420

De Candolle, A. (1883) Origine des Plantes Cultivées. Baillière, Paris

Decraene, L.P.R. & Smets, E.F. (1999) The floral development and anatomy of *Carica papaya* (Caricaceae). *Can. J. Bot.* 77, 582–598

Diaz-Luna, C.L. & Lomeli-Sención, J.A. (1992) Revisión del Género *Jarilla* Rusby (Caricaceae). *Acta Botánica Mex.* 20, 77–99

Farris, D.W., Jaramillo, C., Bayona, G., Restrepo-Moreno, S.A., Montes, C., Cardona, A., Mora, A., Speakman, R.J., Glascock, M.D. & Valencia, V. (2011) Fracturing of the Panamanian isthmus during initial collision with South America. *Geology* 1007–1010

Fay, M.F. & Christenhusz, M.J.M. (2010) Brassicales – an order characterized by shared chemistry. *Curtis's Bot. Mag.* 27, 165–196

Ferrari, L., Orozco-Esquivel, T., Manea, V. & Manea, M. (2012) The dynamic history of the Trans-Mexican Volcanic Belt and the Mexico subduction zone. *Tectonophysics* 522–523, 122–149

Givnish, T.J., Barfuss, M.H.J., Ee, B. Van, Riina, R., Schulte, K., Horres, R., Gonsiska, P.A., Jabaily, R.S., Crayn, D.M., Smith, J.A.C., Winter, K., Brown, G.K., Evans, T.M., Holst, B.K., Luther, H., Till, W., Zizka, G., Berry, P.E. & Sytsma, K.J. (2011) Phylogeny, adaptive radiation, and historical biogeography in Bromeliaceae: Insights from an eight-locus plastid phylogeny. *Am. J. Bot.* 98, 872–895

Godfray, H.C.J. (2002) Challenges for taxonomy. *Nature* 417, 17–19

Godfray, H.C.J. (2007) Linnaeus in the information age. *Nature* 446, 259–60

Gómez-Tuena, A., Orozco-esquivel, M.T. & Ferrari, L. (2007) Igneous petrogenesis of the Trans-Mexican Volcanic Belt. In: Alaniz-Álvarez, S. & Nieto-Samaniego, Á.F. (eds) *Geol. México Celebr. Centen. Geol. Soc. México*, pp. 129–181

Graham, C.H., Ron, S.R., Santos, J.C. & Schneider, C.J. (2004) Integrating phylogenetics and environmental niche models to explore speciation mechanisms in dendrobatid frogs. *Evolution (N. Y)* 58, 1781–1793

Gschwend, A.R., Wai, C.M., Zee, F., Arumuganathan, A.K. & Ming, R. (2013) Genome size variation among sex types in dioecious and trioecious Caricaceae species. *Euphytica* 189, 461–469

Gutiérrez-García, T.A. & Vázquez-Domínguez, E. (2013) Consensus between genes and stones in the biogeographic and evolutionary history of Central America. *Quat. Res.* 79, 311–324

Haug, G.H. & Tiedemann, R. (1998) Effect of the formation of the Isthmus of Panama on Atlantic Ocean thermohaline circulation. *Nature* 393, 1699–1701

Heilborn, O. (1921) Taxonomical and cytological studies of cultivated Ecuadorian species of *Carica*. *Arkiv för Botanik* 17, 1–16

Holstein, N. & Renner, S.S. (2011) A dated phylogeny and collection records reveal repeated biome shifts in the African genus *Coccinia* (Cucurbitaceae). *BMC Evol. Biol.* 11, 28

Hooker, W.J. & Arnott, G.A.W. (1841) The Botany of Captain Beechey's Voyage – comprising an acount of the plants collected by Messrs. Lay and Collie, and other officers of the expedition, during the voyage to the Pacific and Behring's Strait, performed in His Majesty's ship Blossom. H. G. Bohn, London

Hoorn, C., Wesselingh, F.P., ter Steege, H., Bermudez, M.A., Mora, A., Sevink, J., Sanmartín, I., Sanchez-Meseguer, A., Anderson, C.L., Figueiredo, J.P., Jaramillo, C., Riff, D., Negri, F.R., Hooghiemstra, H., Lundberg, J., Stadler, T., Särkinen, T. & Antonelli, A. (2010) Amazonia through time: Andean uplift, climate change, landscape evolution, and biodiversity. *Science* 330, 927–931

Horovitz, S. & Jiménez, H. (1967) Cruzameintos interespecificos intergenericos en Caricaceas ysus implcaciones fitotecnicas. *Agron. Trop.* 17, 323–343

Horovitz, S. & Jiménez, H. (1972) The ambisexual form of *Carica pubescens* Lenné et Koch analized in interspecific. *Agron. Trop.* 22, 475–482

Houle, A. (1999) The origin of platyrrhines: An evaluation of the Antarctic scenario and the floating island model. *Am. J. Phys. Anthropol.* 109, 541–59

Hughes, C. & Eastwood, R. (2006) Island radiation on a continental scale: exceptional rates of plant diversification after uplift of the Andes. *Proc. Natl. Acad. Sci. USA* 103, 10334–10339

IUCN (2013) The IUCN Red List of Threatened Species. Version 2013.2. Available at *www.iucnredlist.org.* Accessed 08 Dec. 2013

Jobin-Decor, M.P., Graham, G.C., Henry, R.J. & Drew, R.A. (1997) RAPD and isozyme analysis of genetic relationships between *Carica papaya* and wild relatives. *Genet. Resour. Crop Evol.* 44, 471–477

Kress, W.J. (2004) Paper floras: how long will they last? A review of Flowering plants of the neotropics. *Am. J. Bot.* 91, 2124–2127

Kumar, L.S.S. & Srinivasan, V.K. (1944) Chromosome number of *Carica dodecaphylla* Vell. *Curr. Sci.* 13, 15

Kyndt, T., Van Droogenbroeck, B., Romeijn-Peeters, E., Romero-Motochi, J.P., Scheldeman, X., Goetghebeur, P., Van Damme, P. & Gheysen, G. (2005a) Molecular phylogeny and evolution of Caricaceae based on rDNA internal transcribed spacers and chloroplast sequence data. *Mol. Phylogenet. Evol.* 37, 442–59

Kyndt, T., Romeijn-Peeters, E., Van Droogenbroeck, B., Romero-Motochi, J.P., Gheysen, G. & Goetghebeur, P. (2005b) Species relationships in the genus *Vasconcellea* (Caricaceae) based on molecular and morphological evidence. *Am. J. Bot.* 92, 1033–1044

Liu, Z., Moore, P.H., Ma, H., Ackerman, C.M., Ragiba, M., Yu, Q., Pearl, H.M., Kim, M.S., Charlton, J.W., Stiles, J.I., Zee, F.T., Paterson, A.H. & Ming, R. (2004) A primitive Y chromosome in papaya marks incipient sex chromosome evolution. *Nature* 427, 348–52

Loera, I., Sosa, V. & Ickert-Bond, S.M. (2012) Diversification in North American arid lands: Niche conservatism, divergence and expansion of habitat explain speciation in the genus *Ephedra. Mol. Phylogenet. Evol.* 65, 437–450

Lorence, D.H. & Colín, R.T. (1988) *Carica cnidoscoloides* (sp.nov) and sect. Holostigma (sect. nov.) of Caricaceae from Southern Mexico. *Syst. Bot.* 13, 107–110

Lysak, M.A., Berr, A., Pecinka, A., Schmidt, R., McBreen, K. & Schubert, I. (2006) Mechanisms of chromosome number reduction in *Arabidopsis thaliana* and related Brassicaceae species. *Proc. Natl. Acad. Sci. USA* 103, 5224–5229

Manshardt, R.M. & Zee, F.T.P. (1994) Papaya germplasm and breeding in Hawaii. *Fruits Var. J.* 48, 146–152

McCormack, J.E., Heled, J., Delaney, K.S., Peterson, A.T. & Knowles, L.L. (2011) Calibrating divergence times on species trees versus gene trees: implications for speciation history of Aphelocoma jays. *Evolution* 65, 184–202

Ming, R., Hou, S., Feng, Y., Yu, Q., Dionne-Laporte, A. et al. (2008) The draft genome of the transgenic tropical fruit tree papaya (Carica papaya Linnaeus). Nature 452, 991–996

Ming, R., Yu, Q. & Moore, P.H. (2007) Sex determination in papaya. Semin. Cell Dev. Biol. 18, 401–408

Montes, C., Cardona, A., McFadden, R., Moron, S.E., Silva, C.A., Restrepo-Moreno, S., Ramirez, D.A., Hoyos, N., Wilson, J., Farris, D., Bayona, G.A., Jaramillo, C.A., Valencia, V., Bryan, J. & Flores, J.A. (2012) Evidence for middle Eocene and younger land emergence in central Panama: Implications for Isthmus closure. Geol. Soc. Am. Bull. 124, 780–799

Müller, B. & Grossniklaus, U. (2010) Model organisms – A historical perspective. J. Proteomics 73, 2054–2063

Olson, M.E. (2002a) Intergeneric relationships within the Caricaceae-Moringaceae clade (Brassicales) and potential morphological synapomorphies of the clade and its families. Int. J. Plant Sci. 163, 51–65

Olson, M.E. (2002b) Combining data from DNA sequences and morphology for a phylogeny of Moringaceae (Brassicales). Syst. Bot. 27, 55–73

Olson, M.E. & Rosell, J.A. (2006) Using heterochrony to detect modularity in the evolution of stem diversity in the plant family Moringaceae. Evolution 60, 724–734

Paterson, A.H., Freeling, M., Tang, H. & Wang, X. (2010) Insights from the comparison of plant genome sequences. Annu. Rev. Plant Biol. 61, 349–372

Penev, L., Kress, W.J., Knapp, S., Li, D.Z. & Renner, S.S. (2010) Fast, linked, and open - the future of taxonomic publishing for plants: launching the journal PhytoKeys. PhytoKeys 14, 1–14

Pennington, R.T., Prado, D.E., Pendry, C.A. & Botanic, R. (2000) Neotropical seasonally dry forests and Quaternary vegetation changes. J. Biogeogr. 27, 261–273

Phillips, S.J. & Dudík, M. (2008) Modeling of species distributions with Maxent: new extensions and a comprehensive evaluation. Ecography 31, 161–175

Pimm, S.L., Russel, G.J., Gittleman, J.L. & Brooks, T.M. (1995) The future of biodiversity. Science 269, 347–350

Pound, M.J., Haywood, A.M., Salzmann, U., Riding, J.B., Lunt, D.J. & Hunter, S.J. (2011) A Tortonian (Late Miocene, 11.61–7.25 Ma) global vegetation reconstruction. Palaeogeogr. Palaeoclimatol. Palaeoecol. 300, 29–45

Prance, G.T. (1984) The pejibaye, Guilielma gasipaes (H.B.K.) Bailey and the papaya, Carica papaya L. In: Stone, D. (ed) Pre-Columbian Plant Migration. Papers Peabody Museum 76, 85–104

Renner, S.S. (2004) Plant dispersal across the tropical Atlantic by wind and sea currents. Int. J. Plant Sci. 165, S23–S33

Scheldeman, X., Willemen, L., Coppens d'Eeckenbrugge, G., Romeijn-Peeters, E., Restrepo, M.T., Romero Motoche, J., Jiménez, D., Lobo, M., Medina, C.I., Reyes, C., Rodríguez, D., Ocampo, J.A., Damme, P. & Goetgebeur, P. (2007) Distribution, diversity and environmental adaptation of highland papayas (*Vasconcellea* spp.) in tropical and subtropical America. *Biodivers. Conserv.* 16, 1867–1884

Schubert, I. & Lysak, M.A. (2011) Interpretation of karyotype evolution should consider chromosome structural constraints. *Trends Genet.* 27, 207–216

Scotland, R.W. & Wood, J.R.I. (2012) Accelerating the pace of taxonomy. *Trends Ecol. Evol.* 27, 415–416

Scotland, R.W. & Wortley, A.H. (2003) How many species of seed plants are there? *Taxon* 52, 101–104

Sepulchre, P., Ramstein, G., Fluteau, F., Schuster, M., Tiercelin, J.J. & Brunet, M. (2006) Tectonic uplift and Eastern Africa aridification. *Science* 313, 1419–1423

Shamir, L., Delaney, J.D., Orlov, N., Eckley, D.M. & Goldberg, I.G. (2010) Pattern recognition software and techniques for biological image analysis. *PLoS Comput. Biol.* 6, e1000974

Silva, E.N. da, Neto, M.F., Pereira, T.N.S. & Pereira, M.G. (2012) Meiotic behavior of wild Caricaceae species potentially suitable for papaya improvement. *Crop Breed. Appl. Biotechnol.* 12, 52–59

Solms-Laubach, H. Graf zu (1889) Die Heimat und der Ursprung des kultivierten Melonenbaumes, *Carica papaya* L. *Bot. Zeitung* 44, 709–720

Stein, L.D. (2008) Towards a cyberinfrastructure for the biological sciences: progress, visions and challenges. *Nat. Rev. Genet.* 9, 678–688

Storey, W.B. (1953) Genetics of the papaya. *J. Hered.* 44, 70–78

The Arabidopsis Genome Initiative (2000) Analysis of the genome sequence of the flowering plant Arabidopsis thaliana. *Nature* 408, 796–815

The Plant List (2010) Version 1. Available at http://www.theplantlist.org/. Accessed 10 Dec. 2013

Tokuhisa, D., Cunha, D., Dos, F., Dias, S., Alvarenga, E.M., Hilst, P.C. & Demuner, A.J. (2007) Compostos fenólicos inibidores da germinação de sementes de mamão (*Carica papaya* L.). *Rev. Bras. Sementes* 29, 161–168

Trauth, M.H., Larrasoaña, J.C. & Mudelsee, M. (2009) Trends, rhythms and events in Plio-Pleistocene African climate. *Quat. Sci. Rev.* 28, 399–411

Van Droogenbroeck, B., Breyne, P., Goetghebeur, P., Romeijn-Peeters, E., Kyndt, T. & Gheysen, G. (2002) AFLP analysis of genetic relationships among papaya and its wild relatives (Caricaceae) from Ecuador. *Theor. Appl. Genet.* 105, 289–297

Van Droogenbroeck, B., Kyndt, T., Maertens, I., Romeijn-Peeters, E., Scheldeman, X., Romero-Motochi, J.P., Van Damme, P., Goetghebeur, P. & Gheysen, G. (2004) Phylogenetic analysis of the highland papayas

(*Vasconcellea*) and allied genera (Caricaceae) using PCR-RFLP. *Theor. Appl. Genet.* 108, 1473–1486

Van Droogenbroeck, B., Kyndt, T., Romeijn-Peeters, E., Van Thuyne, W., Goetghebeur, P., Romero-Motochi, J.P. & Gheysen, G. (2006) Evidence of natural hybridization and introgression between *Vasconcellea* species (Caricaceae) from southern Ecuador revealed by chloroplast, mitochondrial and nuclear DNA markers. *Ann. Bot.* 97, 793–805

Vavilov, N.I. (1987) Origin and Geography Of Cultivated Plants. Translated by D. Löve. Cambridge University Press, Cambridge (1992, English translation)

Vidal, N., Azvolinsky, A., Cruaud, C. & Hedges, S.B. (2008) Origin of tropical American burrowing reptiles by transatlantic rafting. *Biol. Lett.* 4, 115–118

Wang, J., Na, J., Yu, Q., Gschwend, A.R., Han, J. & Zeng, F. (2012) Sequencing papaya X and Yh chromosomes reveals molecular basis of incipient sex chromosome evolution. *Proc. Natl. Acad. Sci. USA* 109, 13710–13715

Wheeler, Q. & Valdecasas, A.G. (2010) Cybertaxonomy and Ecology. *Nat. Educ. Knowl.* 1, 6

Wieczorek, J., Bloom, D., Guralnick, R., Blum, S., Döring, M., Giovanni, R., Robertson, T. & Vieglais, D. (2012) Darwin Core: an evolving community-developed biodiversity data standard. *PLoS One* 7, e29715

Willemse, L.P.M., Welzen, P.C. Van, Mols, J.B., Taxon, S. & May, N. (2008) Standardisation in data-entry across databases: avoiding Babylonian confusion. *Taxon* 57, 343–345

Wilson, E.O. (2003) The encyclopedia of life. *Trends Ecol. Evol.* 18, 77–80

Woodburne, M.O. (2010) The great American biotic Interchange: dispersals, tectonics, climate, sea level and holding pens. *J. Mamm. Evol.* 17, 245–264

Wortley, A.H. & Scotland, R.W. (2004) Synonymy, sampling and seed plant numbers. *Taxon* 53, 478–480

Wu, X., Wang, J., Na, J.K., Yu, Q., Moore, R.C., Zee, F., Huber, S.C. & Ming, R. (2010) The origin of the non-recombining region of sex chromosomes in *Carica* and *Vasconcellea*. *Plant J.* 63, 801–810

Zachos, J., Pagani, M., Sloan, L., Thomas, E. & Billups, K. (2001) Trends, rhythms, and aberrations in global climate 65 Ma to present. *Science* 292, 686–693

Zauner, H. (2009) Evolving e-taxonomy. *BMC Evol. Biol.* 9, 141

Zhang, W., Wang, X., Yu, Q., Ming, R. & Jiang, J. (2008) DNA methylation and hetero-chromatinization in the male-specific region of the primitive Y chromosome of papaya. *Genome Res.* 18, 1938–1943

Appendix

Monograph of Caricaceae

The full taxonomic revision consists of c. 200 pages (Times New Roman, 12 point, single-spaced). Below, I present a few sample pages of the contents on each species, which is freely available in its entirety at http://herbaria.plants.ox.ac.uk/bol/caricaceae

No nomenclatural changes proposed here or in my websites are effectively published (Art. 30 ICBN). Nomenclatural issues aside from those parts already addressed in Chapter III and accepted for publication in *Caricaceae. In: Flora Mesoamericana Vol. 2, Parte 3*, will eventually be formally published according to ICBN rules in peer-reviewed journals of plant taxonomy.

Caricaceae Dumort., Anal. Fam. Pl. 37, 42. 1829, *nom. cons.*, validated by a diagnosis in French; A. DC., Prodr. 15(1): 417. 1864.; V. M. Badillo, Monografía de la família Caricaceae, 222 p. 1971; V. M. Badillo, Alcance 43: 1 – 111. 1993. Type genus: *Carica* L.

Papayaceae Blume, Bijdr.: 940. Jul–Dec 1826, *nom. illeg.*, validated by a diagnosis in Latin. Type genus: *Papaya* Mill., *nom. illeg.* (1754: = *Carica* L., 1753).

Passiflorae tribus Papayaceae Benth. & Hook., Gen. Pl. [Bentham & Hooker f.] 1(3): 809 & 815. 1867.

Herbs, shrubs or trees, with very soft wood. Trunk smooth or spiny, one species endemic to the Sierra de Juarez in Oaxaca, Mexico is completely covered by stinging hairs. Laticifers present, the latex milky, white or yellowish. Leaves alternate in spiral; simple and entire to deeply lobed, or palmately compound. Flowers radially symmetrical, pentamerous, pedicellate or sessile, unisexual (the species dioecious, monecious or polygamous). Male flowers in a panicle, peduncles long or short, small bracts present or absent, calyx small, sepals 5, free or fused basally; corolla of 5 petals, fused in a slender tube, full of nectar; stamens 10, in two whorls of 5 attached to the corolla throat, the superior whorl of stamens alternate with the corolla lobes, the inferior opposite; filaments often free, but sometimes fused at the base forming a short staminal tube; anthers introrse, dehiscing longitudinally, basi- or dorsifixed, mono- or bithecal; connective laminate, often elongated beyond the anther apex, sometimes wider than the anther; pistillode present. Female flowers often solitary or in few-flowered inflorescences; carpels 5, united in a superior ovary, uni- or pentalocular; styles united into a short column, sometimes indistinct; ovules several per carpel, placentae parietal. Fruit a large or small, succulent berry with 1 or 5 cells; seeds numerous, the testa smooth or ornamented with small protuberances or longitudinal ridges; embryo axial; aril fleshy. 34 species and one hybrid within six genera, distributed mostly in the Neotropical region with two species in Africa. Five genera and ten species in Mexico and Central America, one genus with two species in Africa, and two genera and 25 species in South America.

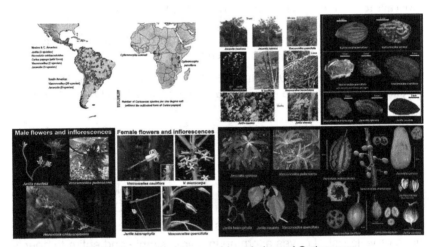

Distribution map and general morphology of Caricaceae

(High-resolution images available at http://herbaria.plants.ox.ac.uk/bol/caricaceae)

Vasconcellea **A. St. –Hil.**, Deux. Mém. Réséd. 12–13. 1837. A. DC., Prodr. 15(1): 415. 1864; Solms, Fl. Bras. 178 –188. 1889; V. M. Badillo, Ernstia 10(2): 74–79. 2000; V. M. Badillo, Ernstia 11(1): 75–76. 2001. Type species: *Vasconcellea quercifolia* A. St. -Hil. Deux. Mém. Réséd. 13. 1837.

Vasconcellea sect. *Hemipapaya* A. DC., Prod. 15(1): 415. 1864. Type species: *Vasconcellea cauliflora* Jacq. (A. DC.).

Vasconcellea sect. *Euvasconcellea* A. DC., Prod. 15(1): 416. 1864. Lectotype species (designated here): *Vasconcellea microcarpa* (Jacq) A. DC.

Carica sect. *Vasconcellea* (A. St.-Hil.) Hooker, Gen. Pl. 1: 815. 1867. Lectotype species: *Carica quercifolia* (A. St.-Hil.) Hieron

Trees or shrubs, stem pithy, smooth, simple or branched. Leaves simple, 1–7 nerved. The number of main veins arising from the base determines the number of lobes, the margins of which may be entire or shallowly to deeply lobed. Flowers white, yellow, or red, greenish or not. Male inflorescence a panicle, congested or with few flowers. Corolla tube glabrous or the throat sparsely pubescent with long and soft hairs. Inferior stamens sessile or subsessile, the connective elongated or not beyond the anther apex. Superior stamens filaments free, the connective rarely elongated beyond anther apex. Ovary 5-locular.

Vasconcellea occurs in South and Central America from Mexico to Uruguay. The highest number of species is found in Southwestern South America more precisely in the Northern Andes region (Ecuador and Peru). Few species are widely distributed along the lowlands.

Notes: The generic name is derived from the name Simão de Vasconcellos a Jesuit who lived in Brazil during the XVII century. *Vasconcellea* was for a long time included as a section (*Vasconcellea*) within the genus *Carica*. Badillo (2000) transferred it again to genus level after the first molecular studies on the family showing that *Vasconcellea* species and *Carica papaya* are not closely related.

Vasconcellea cauliflora (Jacq.) A.DC., Prodr. (DC.) 15(1): 415. 1864. *Carica cauliflora* Jacq., Pl. Hort. Schoenbr. 3: 33-34, t. 311. 1798; *Papaya cauliflora* (Jacq.) Poir., Encycl., Suppl. 4. 296. 1816. Type: Pl. Hort. Schoenbr. 3, t. 311. 1798. (Lectotype designated by Badillo 1993)

Vasconcellea boissieri A.DC., Prodr. (DC.) 15(1): 415. 1864; *Carica boissieri* (A.DC.) Hemsl., Biol. Cent.-Amer., Bot. 1(6): 481. 1880. Type: MEXICO: *Pavón, J.A. s.n.* (holotype G webimage, holotype G webimage, GUADA photo).

Carica bourgeaei Solms, Flora Brasiliensis 13 (3): 178. 1889; *Papaya bourgeaui* (Solms) Kuntze, Revis. Gen. Pl. 1: 253. 1891. Type: MEXICO: Vera Cruz, Córdoba, Vallée du Córdoba, *Bourgeau, E. 2255* (lectotype G webimage, isolectotype F).

Carica quinqueloba Sessé & Moc., Fl. Mexic. (ed. 2). 233. 1894. Type: MEXICO: Puebla, Puebla, Puebla, *Sessé; Mociño Suárez Lozano, J.M. s.n.* (holotype MA, isotype G ! online). MEXICO: *Pavón, J.A. s.n.* (holotype MA, isotype G webimage, GUADA photo from the isotype).

Carica pennata Heilborn, Svensk Bot. Tidskr. 30: 222, fig. 1f, 3. 1936. Type: GUATEMALA: Finca Tiquisate, on the Pacific coast, not far from Rio Bravo Station on the international railway, common at the edge of the tropical forest, Hortus Bergianus 30 Jun 1931 (proveniente de semillas recogidas por N. Johansson), *Johansson, N. 1929* (holotype S webimage, isotypes S webimage, SBT).

Tree 1–4 m tall, dioecious. Petiole 14–37 cm, glabrous. Leaves entire, 5-lobed, apex short acuminate, glabrous on both sides. Male inflorescences growing on the trunk (cauliflorous), peduncle 1–12 cm, thick. Male flowers 35–50 mm; calix 1.5–2 mm, margins entire; corolla tube 25–32 mm; Pistillode 6–7 mm, smaller than 1/2 size of corolla tube, often 1/5 of corolla tube length. Inferior stamens anthers 1.9–2.1 mm; connective elongation 0.9–1.1 mm long (1/2 of the anther length), apex acute or acuminate. Superior stamens filaments 2.3–3 mm, glabrous; anthers 1.4–1.8 mm. Sometimes very sparse and short trichomes are found on the filaments. Female inflorescence (-1) 4–7 flowers, peduncle 1–1.5 cm, thick. Female flowers, pedicel 1.5–2 mm; calix minute (< 1 mm); corolla glabrous, petals 30–37 mm; ovary angled; style 4.8–5.6 mm, conspicuous. Stigmas 6–9 mm, bifurcate. Fruits 2–7 × 4–5 cm, globose, angled, yellow. Seeds 6–9 mm 5–7 mm, oblate; testa cristate with big longitudinal and laminar ribs.

Common names: Chichihua (Mexico), Melocotón (El Salvador), Tapaculo, papayita (Venezuela)

Distribution: It occurs in lowlands wet forests. Sometimes it is also found in cultivation in relatively dry areas, from Southern Mexico to Peru.

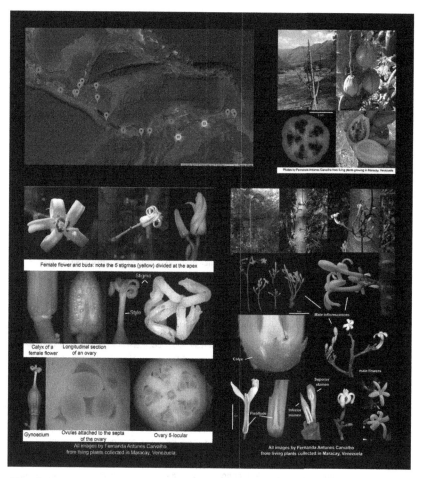

Distribution map based on georeferenced collections and morphology of
Vasconcellea cauliflora

(High-resolution images available at http://herbaria.plants.ox.ac.uk/bol/caricaceae)

Notes: The material distributed by Bourgeau consist of a mix of structures from *Vasconcellea cauliflora* and *Carica papaya*. Badillo (1993) chose the male flowers of a specimen (Bourgeau 2255) in G as the lectotype. An isolectotype in F consists of material exclusively of *Vasconcellea cauliflora* (with fruit and male flowers which might have been collected from different individuals).

Weblinks:

Protologues:

http://www.biodiversitylibrary.org/page/158177#page/421/mode/1up
http://www.botanicus.org/page/158177
http://www.biodiversitylibrary.org/page/272424#page/40/mode/1up
http://www.biodiversitylibrary.org/page/158177#page/421/mode/1up
http://www.botanicus.org/page/158177
http://www.biodiversitylibrary.org/page/142431#page/146/mode/1up
http://www.biodiversitylibrary.org/page/5875934#page/245/mode/1up
http://www.botanicus.org/page/889504

Type specimens:

http://www.ville-ge.ch/musinfo/bd/cjb/chg/adetail.php?id=196520&base=img&lang=en
http://www.ville-ge.ch/musinfo/bd/cjb/chg/adetail.php?id=196520&lang=en
http://www.ville-ge.ch/musinfo/bd/cjb/chg/adetail.php?id=196499&lang=en
http://andor.nrm.se/fmi/xsl/kryptos/fbo/publDetailitems.xsl?-lay=webbkollekter&-token.nav=items&-db=Fbo%20F%C3%96REM%C3%85L&-recid=47168&-find=-find&-token.post=all&-token.languagecode=en-GB
http://www.ville-ge.ch/musinfo/bd/cjb/chg/adetail.php?id=197876&lang=en
http://andor.nrm.se/fmi/xsl/kryptos/fbo/publDetailitems.xsl?-lay=webbkollekter&-token.nav=items&-db=Fbo%20F%C3%96REM%C3%85L&-recid=699233&-find=-find&-token.post=all&-token.languagecode=en-GB

GenBank sequences:

http://www.ncbi.nlm.nih.gov/nuccore/JX092075.1 (Standley, P.C. 89272, Guatemala [F] - ITS)
http://www.ncbi.nlm.nih.gov/nuccore/JX091987.1 (Standley, P.C. 89272, Guatemala [F] psbA-trnH)
http://www.ncbi.nlm.nih.gov/nuccore/JX091939.1 (Standley, P.C. 89272, Guatemala [F] - rbcL)
http://www.ncbi.nlm.nih.gov/nuccore/JX091894.1 (Standley, P.C. 89272, Guatemala [F] rpl20-rps12)
http://www.ncbi.nlm.nih.gov/nuccore/JX092028.1 (Standley, P.C. 89272, Guatemala [F] - matK)
http://www.ncbi.nlm.nih.gov/nuccore/JX091850.1 (Standley, P.C. 89272, Guatemala [F] trnL-trnF)
http://www.ncbi.nlm.nih.gov/nuccore/JX091988.1 (Smith, H.C. 838, Colombia [F] - psbA-trnH)
http://www.ncbi.nlm.nih.gov/nuccore/JX092027.1 (Smith, H.C. 838, Colombia [F] - matK)
http://www.ncbi.nlm.nih.gov/nuccore/JX091938.1 (Romeijn-Peeters, E.H. 284 [GENT])
http://www.ncbi.nlm.nih.gov/nuccore/JX091893.1 (Romeijn-Peeters, E.H. 284 [GENT] rpl20-rps12)
http://www.ncbi.nlm.nih.gov/nuccore/JX092026.1 (Romeijn-Peeters, E.H. 284 [GENT] - matK)
http://www.ncbi.nlm.nih.gov/nuccore/JX091849.1 (Romeijn-Peeters, E.H. 284 [GENT] - trnL-trnF)
http://www.ncbi.nlm.nih.gov/nuccore/DQ061118.1 (Romeijn-Peeters, E.H. 284 [GENT] trnL-trnF)
http://www.ncbi.nlm.nih.gov/nuccore/AY847034.1 (Romeijn-Peeters, E.H. 284 [GENT] trnL-trnF)
http://www.ncbi.nlm.nih.gov/nuccore/AY461561.1 (Romeijn-Peeters, E.H. 284 [GENT] - matK)
http://www.ncbi.nlm.nih.gov/nuccore/AY461546.1 (Romeijn-Peeters, E.H. 284 [GENT] - ITS)

Specimens examined: *Triana 2989* (BM); **BERMUDA**. *Hunter, R. 44* (BM); **COLOMBIA**. *Smith, H.H. 838* (K); *Cuatrecasas, J. 26173* (P); *Triana s.n.* (K); *Losano, G.C. 3909* (VEN); *Lozano, G.C. 3909* (MEXU); **COSTA RICA**. *Vargas, G. 1329* (K); *Soto, R. 4022* (K); *Burger, W.C. s.n.* (BM); *Cascante, A. 1417* (MEXU); *Valverde, O. 200* (K); **EL SALVADOR**. *Rosales, J.M. 726* (BM); *Tucker, J.M. 1321* (K); *Linares, J.L. 12289* (MEXU); **GUATEMALA**. *Friedrichsthal s.n.* (W); *Stevens, W.D. 25530* (MEXU); **HONDURAS**. *Hawkins, T. 984* (MEXU); *Friedrichsthal 1201* (W); **MEXICO**. *Bourgeau, M. s.n.* (K); *Bourgeau, M. 3111* (P); *Goudot, M.J. s.n.* (P); *Miranda, F. 6724* (MEXU); *Matuda, E. 16422* (MEXU); *Calzada, J.I. 9732* (MEXU); *Müller, F. 832* (K,W); *Miranda, F. 7891* (MEXU); *Ortega, J.G. 239* (K); *Ortíz, G.G. 6336* (MEXU); *Vásquez, B. 389* (GUADA); *Castillo, C. 389* (MBM); *Robles, R.G. 698* (GUADA); *Lorence, D.H. 4985* (MEXU); *Rosas, M.R. 849* (BM); **NICARAGUA**. *Moreno, P.P. 21668* (MEXU); **PANAMA**. *Alston, A.H.G. 8875* (BM); *Badillo, V.M. 4155* (MY); *Knapp, S. 1297* (MBM); *Whitefoord, C. 97* (BM); *Knapp, S. 1298* (MEXU); **TRINIDAD AND TOBAGO**. *Brodway, W.E. 7372* (BM); *Comeau, Y.S. 1170* (BM); *Comeau, Y.S. 1237* (BM); **VENEZUELA**. *Pittier, H. s.n.* (VEN); *Jacquin, J. s.n.* (W); *Lister 48* (K); *Ramia, M. 547* (MY); *Schnee, L. 1025* (MY); *Schnee, L. 1041* (MY); *Schnee, L. 1106* (MY); *Cardenas, L. 207* (MY); *E.de Mendiola, B.R.E. 310* (MY); *Edwards, K.S. 413* (K,MY); *Ferrari, G. 111* (MY); *Edwards, K.S. 414* (K); *Agostini, G. 76* (VEN); *Morillo, G.N. 8494* (VEN); *Schnee, L. 1020* (MY); *Iskandar, L. 110* (MY); *Schnee, L. 1111* (MY); *Schnee, L. 1112* (MY); *Schnee, L. 1095* (MY); *Schnee, L. 1105* (MY); *Jaramillo, M. 18* (MY); *Steyermark, J.A. 102199* (MY); *Meier, W. 2575* (MY); *Trujillo, B. 6302* (MY); *Schnee, L. 1013* (MY); *Williams, L. 383* (BM); *López, R.J.L. 750* (VEN); *Ruíz, T. 4144* (MY); *Benítez de Rojas, C.E. 4979* (MY); *Alston, A.H.G. 6990* (BM); *Steyermark, J.A. 19237* (K); *Steyermark, J.A. 96103* (MY); *Delgado, M. 28* (MY); *Berlingeri, C. 113* (MY); *Trujillo, B. 1632* (MY); *Liesner, R.L. 9729* (MY); *Steyermark, J.A. 99562* (MY); *Badillo, V.M. 4518* (MY); *Bunting, G.S. 2773* (MY); *Benítez de Rojas, C.E. 4938* (MY); *Bunting, G.S. 6539* (MY); *Bunting, G.S. 8792* (MY); *Steyermark, J.A. 99561* (MY); *Bunting, G.S. 10188* (MY); *Bunting, G.S. 10195* (MY); *Bunting, G.S. 10207* (MY); *Bunting, G.S. 12362* (MY); *Davidse, G. 18481* (MY); *Davidse, G. 18484* (VEN); *Ijjász, E. 92* (MY); *Gentry, A.H. 41158* (MY); *Badillo, V.M. 4169* (MY); **UNKNOWN**. *Cosson, E. 3869* (P); *HV s.n.* (W); *Hahn, M. 33* (P); *MacLeay s.n.* (K); *Portenschlag-Ledermayer, F. s.n.* (W); *unknown s.n.* (W); *unknown s.n.* (W); *unknown s.n.* (K); *unknown s.n.* (K); *unknown s.n.* (K); *unreadable s.n.bis* (K)

Vasconcellea pubescens A.DC., Prodr. (DC.) 15(1): 419. 1864. *Papaya pubescens* (A.DC.) Kuntze, Revis. Gen. Pl. 1: 253. 1891. Type: PERU: Pasco, Pozuzo, Pozuzo, *Ruiz López, H.; Pavón, J.A. s.n* (Lectotype designated by Badillo (1993) G webimage, isolectotypes F, G webimages [five sheets]).

Carica pubescens Lenne & K.Koch, Index Seminum Hort. Berol. Appendix p. 12. 1854. *Vasconcellea cundinamarcensis* V.M.Badillo, Ernstia 10(2): 78. 2000. Type: ECUADOR: Azuay, Bulán-Paute, Azuay: Bulan-ponte, Siglaloa, Monsica, *Horovitz, S. 1035* (neotype MY, isoneotypes MY, QCA webimage, VEN). Neotype selected by V.M. Badillo (1997).

Vasconcellea cestriflora A.DC., Prodr. (DC.) 15(1): 418. 1864. *Carica cestriflora* (A.DC.) Solms, Flora Brasiliensis 13(3): 185. 1889. *Papaya cestriflora* (A.DC.) & Kuntze, Revis. Gen. Pl. 1: 253. 1891. Type: COLOMBIA: no locality, *Holton 713* (holotype G-BOIS webimage, isotypes G webimage, NY webimage).

Carica candamarcensis Hook.f., Bot. Mag. 101: t. 6198. 1875.; *Papaya cundinamarcensis* (Hook.f.) Kuntze, Revis. Gen. Pl. 1: 253. 1891. Type: ECUADOR: *unknown s.n.* (holotype K, isotypes K, Q).

Carica chiriquensis Woodson, Ann. Missouri Bot. Gard. 45(1): 30-31, f. 6. 1958. Type: PANAMA: Chiriqui, Finca Lerida, al borde quebrada Velo, La Horqueta, cerca de Boquete, *Allen, P.H. 4675* (holotype MO webimage, isotypes G webimage, K).

Tree or shrub 1.5–7 m tall, polygamous. Petiole pubescent. Leaf 5–7-lobed, apex acute to short acuminate; lower size densely pubescent on veins; glabrous above. Male inflorescences axillary, peduncle 4–14 cm, slender. Male flowers 24–27 mm; calix 2–3 mm; corolla tube 12–15 mm; pistillode 5–7 mm, ½ size of corolla tube. Inferior stamens anthers 1.5–2 mm; connective elongation c. 1/3 of the anther length, apex broad to acute. Superior stamens filaments 1.5–2.3 mm long, glabrous or pubescent; anthers 1.3–1.5 mm; connective not elongated. Female inflorescence 4–6 flowers, peduncle 0.7–1 cm. Female flower white or yellow, greenish or not; pedicel 1–4 mm; calix 2–4 mm; corolla glabrous or pubescent outside, petals 20–25 mm; ovary smooth; style indistinct or very short; stigmas 3–7 mm long, apex emarginate or entire. Fruits 6–15 × 3–8 cm, prolate to obovoid, angled. Seeds fusiform 4–5 × 3–3.5 mm, testa with rounded projections.

Common names: Lechoso silvestre (Venezuela), Papaya de montana (Venezuela), Papaya de Ola (Peru)

Distribution: *Vasconcellea pubescens* occurs in cloud forests above 1500 m above sea level in Panama, Colombia, Venezuela, Ecuador, Peru and Bolivia.

Notes: Badillo (1997) selected as the lectotype of *Papaya pubescens* (A.DC.) Kuntze a specimen at G, citing the negative number as F-8515 (which is *Carica glandulosa*), instead of F-8512. However, from the F-8512 negative it is clear that he effectively designated the G specimen that has two attached leaves and 4 inflorescences. The specimen when photographed by the Field Museum had 2 additional loose leaves. Those are not pinned to it anymore and one of the major lobes from one of the leaves has been broken off since the F negative was made. At G the F-8512 negative number is attached to a sheet with a single large leaf, which is not represented in the F photograph. That is also the only one of the 6 specimens that has a G barcode (G00226223). At G one of the specimens has been annotated as a holotype (not the one that Badillo lectotypified) and the rest as isotypes, but that is not original and does not constitute a valid lectotypification. So Badillo's lectotypification is appropriate and his choice (despite the mistake in citing the number) seems clear.

Lenné & K. Koch cite a "specimen a cl. de Warszewicz in Guatemala collectam hortus botanicus ex arboreto regio Sanssouciano (Landesbaumschule) accepit." Neither V. M. Badillo nor I have been able to find a voucher of the plant collected by Warszewicz in Guatemala and cultivated in the Sanssouci garden in Berlin. Badillo neotypified the name with a collection from Ecuador instead of one from Guatemala, the only country mentioned by Lenné & Koch. *V. pubescens,* does not occur in Guatemala; the northern-most occurrence of the species being in Panama. Conceivably, seeds collected somewhere in the Andes were shipped to Europe from Guatemala. Alternatively, the species may have been cultivated in Guatemala. Another possible explanation is that Lenné & Koch pointed Guatemala erroneously as mentioned by V. M. Badillo (1967). Badillo received a letter from Szafer de Cracow saying that there was no *Carica* in the collections notes from Warczewicz. *Carica pubescens* Lenné & Koch may have been collected somewhere in the Andes since Warczewicz spent two years (1851–1853) collecting in that region. The hairy leaves and the 5-locular ovary mentioned by Lenné & K.Koch indicate that the name refers to the morphological species here being treated. When moving *Carica pubescens* Lenné & K. Koch to *Vasconcellea*, Badillo did not cited *Vasconcellea pubescens* A.DC. and thus had to choose a new name because the epithet was occupied by De Candolle's *V. pubescens*. He chose *Carica cundinamarcensis*, based on a *nomen nudum* from Linden's 1871 seed catalogue (price list, item n° 87, p. 52, section on "Arbres et Plantes à Fruits des Tropiques"), which lists *C. cundinamarunsis* (sic) followed by "Des régions froides de Colombie." There is no description and no reference to a specimen. However, the type chosen by Badillo is clearly the same species described by DeCandolle and therefore I fixed in the Flora Mesoamericana the use of *Vasconcellea pubescens* A.DC as the accepted name.

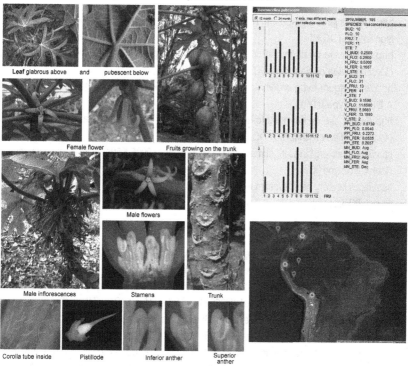

Vasconcellea pubescens A. DC.

Images of living material of *Vasconcellea pubescens* collected by F.A. Carvalho in March 2012 in Merida (Venezuela). The graphs and values (*top right*) show distribution of specimens with flowers, bud and fruits along the year. This data can be used to calculate the Phenological Predictability index that is a tool implemented in BRAHMS and is useful to analyze phenological patterns based on herbarium material (Proença et al. 2012. Ecography 35: 289–293).

Weblinks

Protologues:

http://www.biodiversitylibrary.org/page/158181#page/425/mode/1up

Type specimens:

http://plants.jstor.org/specimen/qca29081 (*Carica pubescens* Lenne & K. Koch., isoneotype [QCA])

http://www.ville-ge.ch/musinfo/bd/cjb/chg/adetail.php?id=196322&lang=en (*Vasconcellea cestriflora* A.DC., holotype [G-BOIS])

http://sweetgum.nybg.org/vh/specimen.php?irn=545025 (*Vasconcellea cestriflora* A.DC., isotype [NY])

http://www.ville-ge.ch/musinfo/bd/cjb/chg/adetail.php?id=196317&lang=en (*Vasconcellea cestriflora* A.DC., isotype [G])

http://apps.kew.org/herbcat/getImage.do?imageBarcode=K000500504 (*Carica candamarcensis* Hook.f., holotype [K])

http://apps.kew.org/herbcat/getImage.do?imageBarcode=K000500503 (*Carica candamarcensis* Hook.f. [K], isotype)

GenBank sequences:

http://www.ncbi.nlm.nih.gov/nuccore/JX092082.1 (Further, H. s.n., Peru [M] - ITS)

http://www.ncbi.nlm.nih.gov/nuccore/JX091996.1 (Further, H. s.n., Peru [M] -psbA-trnH)

http://www.ncbi.nlm.nih.gov/nuccore/JX091906.1 (Further, H. s.n., Peru [M] -rpl20-rps12)

http://www.ncbi.nlm.nih.gov/nuccore/JX092044.1 (Further, H. s.n., Peru [M] -matK)

http://www.ncbi.nlm.nih.gov/nuccore/JX091865.1 (Further, H. s.n., Peru [M] - trnL-trnF)

http://www.ncbi.nlm.nih.gov/nuccore/JX091955.1 (Further, H. s.n., Peru [M] - rbcL)

http://www.ncbi.nlm.nih.gov/nuccore/DQ061117.1 (Romeijn-Peeters E.H. 161 [GENT] - trnL-F)

http://www.ncbi.nlm.nih.gov/nuccore/AY847033.1 (Romeijn-Peeters E.H. 161 [GENT] psbA-trnH)

http://www.ncbi.nlm.nih.gov/nuccore/AY461555.1 (Romeijn-Peeters E.H. 161 [GENT] - matK)

http://www.ncbi.nlm.nih.gov/nuccore/AY461550.1 (Romeijn-Peeters E.H. 161 [GENT] - ITS)

http://www.ncbi.nlm.nih.gov/nuccore/JX091997.1 (HCAR46 [M] - psbA-trnH)

http://www.ncbi.nlm.nih.gov/nuccore/JX091907.1 (HCAR46 [M] - rpl20-rps12)

http://www.ncbi.nlm.nih.gov/nuccore/JX092045.1 (HCAR46 [M] - matK)

http://www.ncbi.nlm.nih.gov/nuccore/JX091866.1 (HCAR46 [M] - trnL-trnF)

http://www.ncbi.nlm.nih.gov/nuccore/JX091956.1 (HCAR46 [M] - rbcL)

Specimens examined: Specimens examined. **BOLIVIA.** *Badillo, V.M. 4069* (MY,MY,VEN); *Badillo, V.M. 4070* (MY); *Badillo, V.M. s.n.* (VEN); *Badillo, V.M. 4042* (MY,VEN); *Badillo, V.M. 4043* (MY,VEN); *Badillo, V.M. 4037* (MY,VEN); *Badillo, V.M. 4038* (MY,VEN); *Badillo, V.M. 4039* (MY,MY,VEN); *Badillo, V.M. 4040* (MY); *Badillo, V.M. 4041* (MY); *Badillo, V.M. 4044* (MY,MY,VEN); *Badillo, V.M. 4047* (MY,VEN); *Badillo, V.M. 4048* (MY,W); *Badillo, V.M. 4049* (MY,VEN); *Badillo, V.M. 4063* (MY); *Badillo, V.M. 4064* (MY,MY,VEN,VEN); **CHILE.** *Schinck, L. 92* (W); **COLOMBIA.**

Holton, I.F. 713 (K); *Ijjász, E. 196* (MY); *Ijjász, E. 198* (MY,MY,VEN); *Correa, J.V. 1580* (BM); *Ijjász, E. 194* (MY,MY,VEN); *Ijjász, E. 195* (MY,VEN); *Sneidern, K. von 1445* (MBM); *Dawe, M.T. 137* (K); *Goudot, M.J. s.n.* (P); *Bursten, B. s.n.* (W); *André, E. 1235* (K); *Cuatrecasas, J. 13645* (MY,P); *Cuatrecasas, J. 13650* (MY,P); *Cuatrecasas, J. 13661* (MY); *Patino, V.M. s.n.* (MY); *Alston, A.H.G. 7735* (BM); *Cuatrecasas, J. 20763* (MY); *Cuatrecasas, J. 18597* (MY); **ECUADOR.** *Palacios, W. 2528* (MY); *Horovitz, S. 1036* (MY); *Horovitz, S. 1027* (MY,MY,VEN); *Horovitz, S. 1022* (MY); *Horovitz, S. 1023* (MY); *Horovitz, S. 1024* (MY,VEN); *Camp 5017E* (MY); *Badillo, V.M. 4238* (MY,VEN); *Alvarez, A. 674* (MY); *Badillo, V.M. 4223* (MY,VEN); *Badillo, V.M. 4239* (MY,VEN); *Badillo, V.M. 4261* (MY,VEN); *Sparre, B. 14580* (P); *Sodiro s.n.* (P); *Badillo, V.M. 4221* (MY,MY,VEN); *Asplund, E. 7426* (R); *Sparre, B. 13968* (GB); *Sodiro s.n.* (P); *Horovitz, S. 1040* (W); *Horovitz, S. 1045* (MY); *Badillo, V.M. 4230* (MY); *Horovitz, S. 1038* (MY); *Horovitz, S. 1039* (MY); *Cornejo, X. 6413* (GB); **HAWAII.** *Degener, O. 34882* (W); **NOVA GRANADA.** *André, E. 457* (K); **PANAMA.** *Arcy, W. d' 13267* (MY); *Badillo, V.M. 4163* (MY,VEN); *Badillo, V.M. 4164* (MY,VEN); **PERU.** *Horovitz, S. s.n.* (VEN); *Martinet, M. s.n.* (P); *Martinet, M. 877295* (P); *Martinet, M. 878295* (P); *Reynel, C. 4151* (K); **VENEZUELA.** *Horovitz, S. 1146* (MY); *Badillo, V.M. 4180* (MY); *Badillo, V.M. 4634* (MY); *Vega, A.V. 37* (MY); *Alston, A.H.G. 6549* (BM); *Trujillo, B. 10897* (MY); *Palacios, S.L. 717* (MY); *Badillo, V.M. 3512* (MY); *Tillet, S.S. 738485* (MY); *Trujillo, B. 6290* (MY); *Mordriz 57* (MY); *Badillo, V.M. 7150* (MY); *Badillo, V.M. 7151* (MY); *Trujillo, B. 24354* (MY); *Badillo, V.M. 7861* (MY); *Gevara, S.C. 879* (MY); *Ijjász, E. 294* (MY); *Steyermark, J.A. 118537* (MY); *Trujillo, B. 8448* (MY); *Ijjász, E. 276 Benítez de Rojas, C.E. 4359* (MY); **UNKNOWN.** *Bonpland, M.A. s.n.* (P); *Hendrickx, F. 6779a* (P); *Hendrickx, F. 6779b* (P); *Rock, J. s.n.* (K).

For images and more details on all specimens cited above, use the link below:
http://herbaria.plants.ox.ac.uk/bol/caricaceae/search?genus=vasconcellea&sp1=pube
scens&exactmatch=false&view=summary

Jacaratia A. DC. Prodr. 15(1): 419. *Jaracatia* Marc., Historiæ rerum naturalium Brasiliæ 8. 1648; V.M. Badillo Monografía de la Familia Caricaceae 206 pp. 1971. Type species: *Carica spinosa* Aubl. (Lectotype designated by Hutchinson 1967).

Pileus Ramírez, Anales Inst. Med.-Nac. Mexico 5(1): 29. 1901. Naturaleza II. 3: 707, pl. 42–45. 1903. Type species: *Pileus heptaphyllus* (Sessé & Moc.) Ramírez.

Leucopremna Standley, Contr. U. S. Natl. Herb. 23(4): 850. 1924. Type species: *Leucopremna mexicana* (A. DC.) Standl.

Trees up to 30 m tall, one species is a shrub. Stem smooth or prickly, branched. Leaves petiolate, palmately compound; leaflets 3–12, sessile or stalked, chartaceous. Seedlings may present lobed leaflets. Male inflorescences axillary; bract minute, caducous or persistent. Male flowers pedicellate, calyx 1–2 mm long; stamens filaments free or shortly fused at the base, the connective not elongated beyond the anthers apex. Female flowers with a short style, often indistinct, stigma entire or digitate at the apex, sometimes bifid. Fruits penduncled, pendent.

Jacaratia digitata (Poepp. & Endl.) Solms-Laub., Flora Brasiliensis 13 (3): 191 (pl. 51, fig. 1.). 1889. *Carica digitata* Poepp. & Endl., Nova Genera ac Species Plantarum 2: 60. 1838.; *Jacaratia spinosa* var. *digitata* (Poepp.) A.DC., Prodr. (DC.) 15(1): 419. 1864. Type: PERU: Maynas, Cochiquinas, *Poeppig, E.F. s.n* (Holotype W, istoypes W, P (2 sheets), L webimage, LZ destroyed).

Jacaratia boliviana Rusby, Bull. New York Bot. Gard. 8(28): 107-108. 1912. Type: BOLIVIA: La Paz, Charopampa, Charopampa, *Williams, R.S. 739* (holotype NY, istotypes US webimage, K, BM).

Tree, perennial, up to 50 m tall. Stem and branches spiny, prickles conical, sharp, up to 5 cm long, latex cream or white. Leaflets 4–7, 9–13 × 3.5–8, oblong to obovate, symmetrical, base acute, apex acuminate; petiole 5–20 cm. Male Inflorescences dense (many flowers), peduncle 4–15 cm. Male flowers calyx glabrous, the margin entire or erose (herbarium specimens present lighter margins); corolla externally green, internally white; tube 8.5–13 mm, glabrous inside, the pistillode c. ½ of the corolla tube length. Inferior stamens pedicellate, filaments c. 1.8 mm, glabrous, the connective not elongate, the anthers 3–4 mm. Superior stamens filaments 3–4 mm, partially fused to the filaments of the inferior stamens, the connective glabrous, not elongated, the anthers 2–2.5 mm. Female Inflorescences with 1–2 flowers, peduncle 2–4.5 cm. Female flowers light yellow, greenish or not; calyx c. 1 mm, the margins entire; corolla with petals 15–23 mm; ovary costate with 5 angles; stigmas 4.5–6.5 mm, apex entire. Fruits spindle-shaped to ellipsoid, 10–15 cm long, yellow to bright orange, smooth or sometimes with maroon stripes, many seeded. Seeds c. 6 × 8 mm with laminar projections.

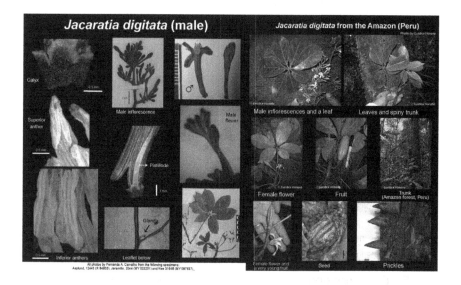

Jacaratia digitata (male)

Calyx

Superior anther

Male inflorescence

Pistillode

Inferior anthers

Gland

Leaflet below

All photos by Fernanda A. Carvalho from the following specimens:
Asplund, 12443 (R 84653), Jaramillo, 2044 (MY 023201) and Nee 31648 (MY 087937)

Jacaratia digitata from the Amazon (Peru)

Male inflorescences and a leaf

Leaves and spiny trunk

Female flower

Fruit

Trunk (Amazon forest, Peru)

Female flower and a very young fruit

Seed

Prickles

Male flower

Pollination: Male flowers are reported as having a penetrating unpleasant chlorine odor and visited by heliconid butterflies (specimen M. Nee 39232).

Phenology: Fertile specimens were collected along the whole year.

Common names: Assacú-Branco, Rihuwarisi, Mamaozinho, Mamuim, Jaracata, Jaracatiá, Mamu-í, Mamaui (Brazil), Gargatea (uruna, Bolivia), Papaya de monte, Papayllo, papayón (Bolivia), Papaya caspi, Shamburo (Peru), yuquilla, Papayillo, Numpi (Ecuador).

Distribution: *Jacaratia digitata* occurs in the western Amazonia, from west Pará (Brazil) to Bolivia, Peru, Ecuador and Peru. It grows in the Amazon forest in clayey and sandy soils along non- flooded forests ("terra firme" forests).

Taxonomic notes: The name *Jacaratia* is derived from the Tupi Iacaratiá, still used today by local people in Brazil. De Candolle probably misspelled the name originally published by Georg Marcgrave in 1648 as *Jaracatia*. Poeppig and Endlicher (1838) wrote "*C. digitata* Aubl. Guyan II. 908. Tab 346. *C. spinosa* Willd. Pers.". However Aublet (1775) described and illustrated *Carica spinosa* a synonym of *Jacaratia spinosa* (Aubl.) A.DC. Therefore, *C. digitata* Aubl. is invalid.

Weblinks

Protologues:

http://www.biodiversitylibrary.org/page/142434#page/153/mode/1up
http://www.botanicus.org/page/142437

Types:

http://apps.kew.org/herbcat/getImage.do?imageBarcode=K000500513 (*Jacaratia digitata* (Poepp. & Endl.) Solms-Laub. [istotype K])

http://collections.mnh.si.edu/search/botany/ (*Jacaratia boliviana* Rusby [istotype US])

http://vstbol.leidenuniv.nl/NHN/image/L0010313_HERB.jpg (*Carica digitata* Poepp. & Endl. [istoype L])

GenBank sequences

http://www.ncbi.nlm.nih.gov/nuccore/JX092057.1 (Monteagudo, A. 19254, Ecuador [LOJA] - ITS)
http://www.ncbi.nlm.nih.gov/nuccore/JX091919.1 (Romeijn-Peeters, E.H. 36, Ecuador [GENT] rbcL)
http://www.ncbi.nlm.nih.gov/nuccore/JX091880.1 (Romeijn-Peeters, E.H. 36, Ecuador [GENT] rpl20-rps12)
http://www.ncbi.nlm.nih.gov/nuccore/JX092009.1 (Romeijn-Peeters, E.H. 36, Ecuador [GENT] matK)
http://www.ncbi.nlm.nih.gov/nuccore/JX091831.1 (Monteagudo, A. 19254, Ecuador [LOJA] trnL-trnF)
http://www.ncbi.nlm.nih.gov/nuccore/JX091830.1 (Romeijn-Peeters, E.H. 36, Ecuador [GENT] trnL)
http://www.ncbi.nlm.nih.gov/nuccore/DQ061138.1 (Romeijn-Peeters, E.H. 36, Ecuador [GENT] trnL-trnF)
http://www.ncbi.nlm.nih.gov/nuccore/AY461574.1 (Romeijn-Peeters, E.H. 36, Ecuador [GENT] matK)

Specimens examined: **BOLIVIA**. *Smith, D.N. 13003* (MY); *Wood, J.R.I. 12787* (K,LPB); *Smith, D.N. 12847* (LPB,MEXU,MY); *Badillo, V.M. 4058* (MY,VEN); *unreadable 1877* (K); *Cruz, A. 1* (K); *Seidel, R. 9106* (MY); *Araujo, A.M. 3123* (MY); *Badillo, V.M. 4059* (MY); *Badillo, V.M. 4060* (MY); *Nee, M. 31648* (MY); *Nee, M. 31651* (LPB,MY); *Pennington, T.D. 18* (K); *Pennington, T.D. 38* (K); *Nee, M. 36883* (MY); *Nee, M. 37241* (MY); *Nee, M. 39080* (MY); *Nee, M. 39232* (MY); *Guillén, R. 3030* (MY); *Guillén, R. 3531* (MY); *Guillén, R. 3984* (MY); *Rodríguez, A. 736* (MY); **BRAZIL**. *Sasaki, D. 1491* (K); *Sasaki, D. 222* (K); *Zappi, D. 1431* (K); *Krukoff, B.A. 5446* (BM,K); *Daly, D.C. 9499* (MY); *Daly, D.C. 7452* (MY); *Ferreira, C.A.C. 10455* (INPA,MY); *Ferreira, C.A.C. 10763* (INPA,MY); *Prance, G.T. 12217* (INPA,K,MY); *Daly, D.C. 7678* (MY); *Figueiredo, C. 562* (W); *Nelson, B.W. 815* (INPA,K); *Paula, A. 62* (INPA); *Souza, J.M.A. de s.n.* (INPA);

Vasconcellos, D. s.n. (INPA); *Daly, D.C. 7881* (MEXU,MY); *Daly, D.C. 8054* (INPA); *Prance, G.T. 7816* (INPA,K,MY,R); *Ramos, J.F. 671* (INPA); *Souza, J.M.A. de s.n.* (INPA); *Daly, D.C. 8726* (MY); *Silveira, M. 820* (MY); *Daly, D.C. 7286* (INPA,MY); *Krukoff, B.A. 4791* (K); *Milliken, W. 1891* (INPA,K); *Ule, E. 966* (K); *Luize, B.G. 416* (INPA); *Ducke, A. 1881* (IAN,K,R); *Krukoff, B.A. 6352* (BM,K); *Krukoff, B.A. 8183* (BM,K,P); *Rosa, N.A. 2171* (INPA,MBM,RB); *Roth, P. 02* (INPA); *Sobral, M. 9919* (BHCB); *Bilby, R. 122* (INPA); *Rodrigues, W.A. 4332* (INPA); *Rodrigues, W.A. 9647* (INPA); *Silva, J.A. da 59* (INPA); *Vieira, G. 379* (MY,RB); *Vieira, M.G.G. 379* (INPA); *Ferreira, C.A.C. 8706* (INPA,K,MY); *Ferero, E. 7116* (MY); *Forero, E. 7116* (F,K,R); *Ferreira, C.A.C. 4858* (INPA,K,RB); **COLOMBIA**. *Schultes, R.E. 8318* (K); *Jaramillo, J.R.M. 2044* (MY); *Callejas, R. 6038* (MY); *Philipson, W.R. 1584* (BM); *Philipson, W.R. 1908* (BM); **ECUADOR**. *Villa, G. 1005* (BM); *Neill, D.A. 7469* (MY); *Zaruma, J. 477* (K,MY); *Aulestia, M. 1178* (MY); *Vargas, H. 903* (MY); *Neill, D.A. 6277* (MY); *Palacios, W. 806* (MY); *Zaruma, J. 128* (MY); *Palacios, W. 2767* (GB,K); *Palacios, W. 313* (MY); *Cerón, C.E. 1262* (MY); *Neill, D.A. 7093* (MY); *Zaruma, J. 766* (GB,MY); *Miller, J.S. 755* (MY); *Clark, J.L. 1208* (MY); *Pennington, T.D. 10752* (K); *Badillo, V.M. 4233* (BM,MY,P,RB); **PERU**. *Jaramillo, M. 2074* (MY); *Schunke-Vigo, J. 3933* (IAN); *Wurdack, J.J. 2021* (P); *Huashikat, V. 611* (MY); *Acevedo-Rodriguez, P. 8889* (K,P); *Honorio, E. 1300* (MOL); *Honorio, E. 1301* (M,MOL); *Honorio, E. 1306* (MOL); *Honorio, E. 1317* (MOL); *Honorio, E. 1336* (MOL); *Honorio, E. 1339* (MOL); *Honorio, E. 1340* (MOL); *Honorio, E. 1343* (MOL); *Honorio, E. 1346* (MOL); *Asplund, E. 12443* (R); *Honorio, E. 1242* (M,MOL); *Honorio, E. 1243* (MOL); *Honorio, E. 1247* (MOL); *Honorio, E. 1252* (MOL); *Honorio, E. 1253* (M,MOL); *Honorio, E. 1255* (MOL); *Honorio, E. 1257* (MOL); *Honorio, E. 1260* (MOL); *Honorio, E. 1264* (M,MOL); *Honorio, E. 1269* (MOL); *Honorio, E. 1281* (M,MOL); *Foster, R.B. 9345* (MY); *Badillo, V.M. 4693* (MY); *Pennington, T.D. 16523* (K); *Gentry, A.H. 26935* (MY); *Gentry, A.H. 28261* (MY); *Asplund, E. 14468* (K); *Ayala, F. 2199* (MY); *Honorio, E. 1003* (M,MOL); *Honorio, E. 1012* (MOL); *Honorio, E. 1021* (MOL); *Honorio, E. 1029* (MOL); *Honorio, E. 1030* (MOL); *Honorio, E. 1036* (MOL); *Honorio, E. 1041* (MOL); *Honorio, E. 1047* (MOL); *Honorio, E. 1049* (M,MOL); *Honorio, E. 1000* (MOL); *Ruiz, J. 1196* (K); *Ruíz, J. 3396* (MY); *Vasquez, R. 1314* (MY); *Vásquez, R. 6868* (K,MEXU); *Honorio, E. 1059* (MOL); *Honorio, E. 1077* (MOL); *Honorio, E. 1082* (MOL); *Honorio, E. 1093* (MOL); *Honorio, E. 1105* (MOL); *Honorio, E. 1109* (MOL); *Honorio, E. 1111* (MOL); *Honorio, E. 1114* (M,MOL); *Honorio, E. 1116* (MOL); *Gentry, A. 27032* (MY); *Ule, E. 9646* (RB); *Alexiades, M. 359* (MY); *Alexiades, M. 381* (MY); *Gentry, A.H. 45620* (MY); *Gentry, A.H. 45769* (MEXU,MY); *Monteagudo, A. 12644* (HOXA); *Reynel, C. 5231* (MY); *Monteagudo, A. 15876* (HOXA); *Wallnöfer, B. 2328815* (K,W); *Wallnöfer, B. 8118813* (K,W); *Souza,*

J. de 112 (INPA,MY); *Klug, G. 3816* (BM,K); *Boeke, J.D. 1277* (MY); *Schunke, J.V. 3933* (K,P); *Knapp, S. 7454* (MEXU,MY,P); *Schunke, J.V. 14758* (MY); *Tello 426* (K); *Honorio, E. 1202* (MOL); *Honorio, E. 1203* (MOL); *Honorio, E. 1204* (MOL); *Honorio, E. 1209* (M,MOL); *Honorio, E. 1216* (MOL); *Honorio, E. 1217* (M,MOL); *Honorio, E. 1222* (MOL); *Honorio, E. 1227* (MOL); *Honorio, E. 1229* (MOL); *Honorio, E. 1234* (MOL); UNKNOWN. *White, G.E. 1064* (K).

For images and more details on all specimens cited above, use the link below:
http://herbaria.plants.ox.ac.uk/bol/caricaceae/search?genus=jacaratia&sp1=digitata&
exactmatch=false&view=summary

Jarilla Rusby, Torreya 21: 47. 1921. *Mocinna* Cerv. ex La Llave, Reg. Trim. 1: 351. 1832. Homonym of *Mocinna* Lag. Gen. Sp. Pl. 31. 1816, a genus of Asteraceae. Type species: *Jarilla heterophylla* (Cerv. Ex. La Llave) I. M. Johnst

Dioecious herbs. Stems annual, branched. Tubers perennial, fusiform or globose. Leaves simple, 3–5 lobed, never deeply lobed. Male inflorescence axillary, often long peduncled. Calyx 5 lobed united at the base in a short tube; aestivation imbricate, corolla infundibular, white or purplish-white; stamens united at the base forming a short tube, rarely free; filaments pilose; anthers of superior stamens short and monothecal; anthers of inferior stamens large and bithecal; connectives pubescent, not elongated beyond the anthers apex. Female flowers, often solitary or in small cymes of 2–3, pedicellate. Ovary unilocular; style indistinct. Fruit a berry; pendent, mucronate, smooth or with longitudinal wings. Seeds numerous, testa smooth or slightly tuberculate.

Three species occurring from northwestern Mexico to Guatemala.

Jarilla chocola Standl., Publ. Field Mus. Nat. Hist., Bot. Ser. 17: 200. 1937. Type: MEXICO: Sonora, Chihuahua, Guasarema, Rio Mayo, *Gentry, H.S. 2366* (holotype F, isotypes F, GUADA, GUADA, K, S, US n.v.). MEXICO: Sonora, San Bernardo, Rio Mayo, *Gentry, H.S. 1624* (syntypes F, GUADA, MEXU).

Jarilla heterophylla (Cerv. ex La Llave) Rusby ssp. *pilosa* V.M.Badillo, Monogr. Fam. Caricaceae, 163.1971. Type: GUATEMALA: Jutiapa, Moyuta, Moyuta, *Morales, J. s.n.* (holotype US *n.v.*).

Herbs erect, mostly unbranched, up to 1 m tall. Stem succulent, glabrous or rarely puberulent. Tuber fusiform 5–15 cm long, generally slender, 1–5 cm diameter. Blade deltoid to ovoid, entire to 1–5 lobed; margins entire; base truncate, rounded or sub-cordate; apex acuminate. Male inflorescence peduncle 4–10 cm. Male flowers 6.5–9.3 mm, white, sometimes with purple longitudinal stripes; calix 0.5–0.8 mm long; corolla tube 3.4–5 mm; pistillode 2–3.5 mm, longer than ½ size of corolla tube length. Inferior stamens filaments 0.2–0.4 mm, pubescent; anthers 0.8–1 mm. Superior stamens filaments 0.5–0.7 mm; anthers 0.7–1 mm. Female inflorescence 1–2 flowers, peduncle 0.3–0.7 cm. Female flowers pedicel 1.8–2.8 mm; calyx 0.6–1 mm, glabrous, apex obtuse or acute; corolla 5.5–8 mm, apex rounded to acute; ovary pyriform, with 5 longitudinal wings, each wing with a basal appendage; stigma 2–3 mm densely pubescent. Fruits 2–7 1.5–4 cm, solitary, pink to dark brown, ovoid, base truncate or slightly concave, apex tapered; peduncle 1–7 cm; longitudinal wings are prolonged at the base into falcate appendages. Seeds 3.5–4.5 × 2.5–3.5 mm, black or dark brown; testa smooth sometimes slightly ruminate.

Common names: Jarrilla, Jarrilla de coyote, Granadilla, Machicua, Chivitos, Huevitos de Venado, Toritos

Distribution: It occurs in seasonally dry forests along the Pacific Coast from Northwestern Mexico, to El Salvador.

Phenology: Leaves appears initially in June. Flowers from June to August and fruits until October.

Notes: A papain-like proteinase is reported for tubers and fruits (Tookey and Gentry 1969). However cultivation can be problematic due to freezing intolerance and susceptibility to soil parasites (Willingham B.C. and White G.A. 1976)

Jarilla chocola

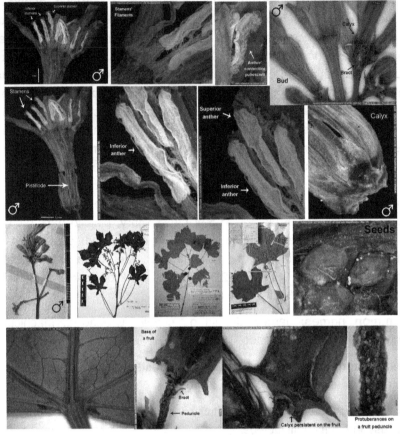

All images from the following herbarium specimens: Gentry 1624 (K, paratype), Gentry 2366 (F, K, types), Gentry 1624 (K), Reina 97578 (Mexu) by Fernanda Antunes Carvalho

Weblinks

Living plants:

http://www.desertmuseum.org/programs/yecora_gallery.php

Type specimens:

http://apps.kew.org/herbcat/detailsQuery.do?imageId=202389&pageCode=1&presentPage=1&queryId=2&sessionId=CA72521AFB52A76EAFF3F5392755D99B&barcode=K000500520 (Isotype of *Jarilla chocola* Standl. [K])

http://andor.nrm.se/fmi/xsl/kryptos/fbo/publDetailitems.xsl?-lay=webbkollekter&-
token.nav=items&-db=Fbo%20F%C3%96REM%C3%85L&-recid=18204&-find=-find&-
token.post=all&-token.languagecode=en-GB (Isotype of *Jarilla chocola* Standl. [S])

GenBank sequences:

http://www.ncbi.nlm.nih.gov/nuccore/JX092064.1 (Lott, E.J., Mexico 31 [MEXU] - ITS region)
http://www.ncbi.nlm.nih.gov/nuccore/JX091977.1 (Lott, E.J., Mexico 31 [MEXU] *−psbA-trn*H)
http://www.ncbi.nlm.nih.gov/nuccore/JX091927.1 (Lott, E.J., Mexico 31 [MEXU] - *rbc*L)
http://www.ncbi.nlm.nih.gov/nuccore/JX091884.1 (Lott, E.J., Mexico 31 [MEXU] - *rpl*20-*rps*12)
http://www.ncbi.nlm.nih.gov/nuccore/JX092017.1 (Lott, E.J., Mexico 31 [MEXU] - *mat*K)
http://www.ncbi.nlm.nih.gov/nuccore/JX091838.1 (Lott, E.J., Mexico 31 [MEXU] - *trn*L-F)
http://www.ncbi.nlm.nih.gov/nuccore/AF378624.1 (Reina 99-962A [MO] - *trn*G)
http://www.ncbi.nlm.nih.gov/nuccore/AF378577.1 (Reina 99-962A [MO] - ITS region)
http://www.ncbi.nlm.nih.gov/nuccore/JX091976.1 (Gentry, H.S. 1553, Mexico [F] *-psbA-trn*H)
http://www.ncbi.nlm.nih.gov/nuccore/JX091837.1 (Gentry, H.S. 1553, Mexico [F] - trnL-Leu)

Specimens examined. EL SALVADOR. *Linares, C.A. 5001* (MEXU);
Linares, J.L. 5000 (MEXU); *Linares, J.L. 7554* (MEXU); *Linares, J.L. 7555*
(MEXU); *Linares, J.L. 7559* (MEXU); **MEXICO.** *Cruz, A.L. 1289* (MEXU);
Palacios, E.E. 745 (GUADA); *Miranda, F. 5291* (MEXU); *Vázquez, M.A.D.
376* (MEXU); *Gentry, H.S. 1553* (F); *McVaugh, R. 15762* (MEXU); *Vazquez,
L.V. 878* (MEXU); *Cházaro, M.B. 4837* (MEXU); *Santana, F.J.M. 7925*
(MEXU); *Lott, E.J. 31* (MEXU); *Lott, E.J. 1422* (MEXU); *Magallanes, J.A.S.
3815* (MEXU); *Magallanes, J.A.S. 4256* (MEXU); *Wilbur, C.R. 1594* (MEXU);
Pérez, L.A.J. 1765 (MEXU); *Harker, M. 11194* (IBUG); *Chávez, S.T. s.n.*
(GUADA); *Luna, C.L.D. 20832* (GUADA); *Sención, J.A.L. s.n.* (GUADA);
Carvalho, F.A. 2309 (MEXU); *Gentry, H.S. 10753* (MEXU); *Luna, C.L.D.
20819* (GUADA); *Zafra, A.N. 1068* (MEXU); *Colín, R.T. 13783* (MEXU);
Sousa, M. 763 (MEXU); *Hernandez, F.A. 848* (MEXU); *Reina, A.L. 97578*
(MEXU); *Van Devender, T.R. 99336* (MEXU); *Búrquez, A. 96658* (MEXU).

For images and more details on all specimens cited above, use the link below:
http://herbaria.plants.ox.ac.uk/bol/caricaceae/search?genus=jarilla&sp1=chocola&exa
ctmatch=false&view=summary

Acknowledgments

Many thanks go to Susanne Renner for her advice and unconditional support throughout these years. She always found time to discuss my ideas, observe my progress, listen to my sorrows, and to read my long manuscripts, including the annual reports to CNPq in Portuguese on the 31st of December each year. Dear Susanne, thanks for sharing with me all your enthusiasm and broad knowledge, for helping me to improve my writing skills and my oral presentations. This work would be impossible without your great supervision.

Martina Silber always found solutions for the weirdest laboratory problems, and provided emotional support in many difficult moments. Martina, vielen herzlichen Dank!

Thanks also to Emilie Vosyka who kindly helped and taught me the laboratory work during the initial phase of my Ph.D. when I had my first contact with high-tech pipettes. Mila, your help was invaluable. Vielen Dank!

Harald Loose, Andreas Richter, and Achim Hörmann taught me the basic knowledge for cultivating plants, and helped me to keep my lovely Caricaceae growing healthy in the greenhouses of the Botanical Garden.

Fieldwork in Mexico was only possible because of the collaboration with José Lomeli who showed me the tasty *Jarrillas y toritos* from Jalisco and Nayarit, and introduced me to the great flavor of coconut with pepper. Susanne Magallón introduced me to Pedro Tenorio and Rafael Torres with whom I formed the great team of "Horovitzios" in the mountains of Sierra de Juaréz. Muchas gracias Rafa and Pedro for the wonderful time looking for the Horovitzias!

Thanks to Gilberto Morillo for introducing me to the Venezuelan Andes in Merida (Venezuela).

Carmen Emília Benítez and Aurimar Magallanes were not only extremely helpful, but also nice friends during my long stay in Maracay (Venezuela). I am also very grateful to the whole team of the library and the herbarium Victor Badillo, who provided integral assistance during the five weeks I spent working in Maracay.

Many thanks go to the curators of the herbaria for giving me access to the collections and great assistance.

My dear fellows from the Botanical Institute are commended for helping whenever help was needed, especially Juliana Chácon, Sidonie Bellot, Aretuza Sousa, Aline Martins, Patrizia Sebastian, Natalia Filipowizc, Florian Jabbour, Natalie Cusimano, Stefan Abrahamczyk, Marc Gottschling, Oscar Pérez, Mathieu Piednoël, Christian Brauchler, Günter Gerlach, Lars Nauheimer, Norbert Holstein, Daniel Souto, Eva Facher, Martina Simbeck, and Andreas Groeger.

Special thanks to the lovely friends I met in Munich and became an important and memorable part of my life here, and those from Brazil with whom I managed to keep contact throughout the cyberspace sharing moments of anguish and achievements even so far apart.

I am indebted to Theodor C. H. Cole (Heidelberg) who not only helped me with learning English since my first days in Heidelberg, and who is currently proofreading and editing my e-Monograph, but also became a great friend in moments of happiness or difficulties.

My deep love and appreciation goes out to my family: to my parents, Bia and Quincas, and my sister, Tânia, who always supported me and provided essential encouragement even from the distance – and to my husband, Alexandre, who has been a reliable and brilliant companion, an inspiring fellow adventurer, encouraging me in all moments of my life during the past 12 years. Mãe, Pai, Tânia e Ale muito obrigada por estarem sempre por perto!

Curriculum Vitae

Name	Fernanda Antunes Carvalho
Address	Systematische Botanik
	Menzinger Str. 67
	80638 München, Deutschland
	Phone: 0049 01724705793
	Email: antunesfc@gmail.com
Born	13 May 1981, Belo Horizonte, MG, Brazil

2010–2014   Ph.D. candidate in Ecology, Evolution, Systematics
Ludwig-Maximilians-Universität, München
Department für Biologie
Systematische Botanik und Mykologie
Dissertation: "Molecular phylogeny, biogeography, and
an e-monograph of the papaya family (Caricaceae)"
Advisor: Prof. Dr. Susanne S. Renner

2004–2006   Master in Botany
(Master Grant Holder – Brazilian Government/CNPq)
National Institute of Amazon Research – INPA
Thesis: "Beta diversity between Purus and Madeira
rivers: determinants of the structure of communities of
Marantaceae, Araceae and Pteridophytes in BR 319,
Amazonas, Brazil"
Advisor: Dr. Flávia R. C. Costa

2001–2004   Diploma in Biological Sciences
Federal University of Minas Gerais, UFMG, Brazil
Advisor: Dr. Alexandre Salino

Portuguese (native), English (fluent), Spanish (good), German (basic)

Professional Experience

Feb-Mar 2009   Research Assistant at the laboratory of
                Prof. Dr. Susanne S. Renner, Faculty of Biology
                (Ludwig Maximilians University, Munich) Holder: STIBET/DAAD

Participation in projects

2006–2009   "Biodiversity survey in savanna areas between the
                Madeira and Purus rivers, Amazonas, Brazil".
                Holder: Brazilian Government – GEOMA Network

2006–2008   "Structure, biomass and composition of trees and
                herbaceous species of the Biological Reserve of Uatumã,
                Amazonas, Brazil" (Brazilian Government Scholarship –CNPq).
                Holder: Brazilian Government – CNPq/PPG7/MCT

2004–2006   "Study of beta diversity of a plant community over
                interfluve Purus-Madeira to determine priority areas for
                conservation of the Amazon." (Brazilian Government
                Scholarship – CNPq). Holders: Federal Brazilian Government –
                CNPq; State Government of Amazonas – FAPEAM

2004   "Pteridophytes survey at Neblina Mountains, Amazonas
                state, Brazil" Holder: Brazilian Institute of Environment
                and Renewable Natural Resources – IBAMA

2003   "Biological Inventory in the basin of Jequitinhonha and
                Mucuri rivers in the states of Minas Gerais and Bahia,
                Brazil" (Scientific Initiation Grant – CNPq) Holders:
                NPq/PROBIO/MMA/Conservation Intl-Brazil, Atlantic Forest Program

2003–2004   "Floristic survey at Parque Natural Ribeirão do Campo,
                Espinhaço Range, Minas Gerais, Brazil" (Field Research
                Assistant) Holder: Department of Environment of Conceição do
                Mato Dentro

2003	"Community, Reproductive Biology and Diet of Birds in four Forest Environments of the Pantanal de Poconé, MT, Brazil" (Field Research Assistant). Volunteer (35 days)
2001–2002	"Plant Diversity Conservation of the Doce river basin in Minas Gerais: Study floristic and phytosociological" (Scientific Initiation Grant – CNPq). Holder: CNPq/PELD/UFMG
2001–2002	"Diversity of Pteridophytes in conservation areas of the Atlantic Forest of the state of Sao Paulo, Brazil" (Field Research Assistant). Holder: O Boticário Fundation

Invited presentations and workshops

Introduction to BRAHMS
Department of Biological and Environmental Sciences, University of Gothenburg, One day workshop. Gothenburg, Sweden. 19 Sep. 2013

Biogeography of Caricaceae, and a cyber-monograph of the papaya family
Department of Biological and Environmental Sciences,
Univ. of Gothenburg, Invited talk. Gothenburg, Sweden. 15 Apr. 2013

Evolution and biogeography of Caricaceae, and the closest relatives of papaya
Instituto de Biología de la Universidad Nacional Autónoma de México. Invited talk. Mexico City, Mexico. 7 Aug. 2012

Una introdución sobre BRAHMS (Botanical Research and Herbarium Management System)
Insituto de Botánica Agrícola, Universidad Central de Venezuela.
Invited talk. Maracay, Venezuela. 26 Mar. 2012

Publications

Peer-reviewed journal articles

Carvalho F.A., Filer D., Renner S.S. (2014) Taxonomy in the electronic age and an e-monograph of the papaya family (Caricaceae) as an example. *Cladistics* (publ. online 13 Aug. 2014), doi:10.1111/cla.12095

Carvalho F.A. & Renner S.S. (2013) Correct names for some of the closest relatives of *Carica papaya*: A review of the Mexican/ Guatemalan genera *Jarilla* and *Horovitzia*. *PhytoKeys* 29: 63–74. doi:10.3897/phytokeys.29.6103

Pansonato, M.P., Costa, F.R.C., de Castilho, C.V., **Carvalho, F.A.**, Zuquim, G. (2013) Spatial scale or amplitude of predictors as determinants of the relative importance of environmental factors to plant community structure. *Biotropica* 45: 299–307. doi: 10.1111/btp.12008

Carvalho F.A. & Renner, S.S. (2012) A dated phylogeny of the papaya family (Caricaceae) reveals the crop's closest relatives and the family's biogeographic history. *Molecular Phylogenetics and Evolution* 65(1): 46–53

Carvalho, F.A., Salino, A., Zartman, C.E. (2012) New Country and Regional Records from the Brazilian Side of Neblina Massif. *American Fern Journal* 102: 228–232

Pezzini F.F., Melo, P.H.A., Oliveira, D.M.S., Amorim, R.X., Figueiredo, F.O.G., Drucker, D., Rodrigues, F.R.O., Zuquim, G.P.S, Sousa, T.E.L., Costa, F., Magnusson, W.E., Sampaio, A.F., Lima, A.P., Garcia, A.R.M., Manzatto, A.G., Nogueira, A., Costa, C.P., Barbosa, C.E.A., Castilho, C.V., Cunha, C.N., Freitas,C.G., Cavalcante, C.O., Brandao, D., Rodrigues, D. J., Santos, E.C.P.R., Baccaro, F.B., Ishida, F.Y., **Carvalho, F.A.**, Moulatlet, G.M., Guillaumet, J.L.B., Pinto, J.L.P.V., Schietti, J., Vale, J.D., Belger, L., Verdade, L.M., Pansonato, M.P., Nascimento, M.T., Santos, M.C.V., Cunha, M.S., Arruda, R., Barbosa, R.I., Romero, R.L., Pansini, S., Pimentel, T.P. (2012) The Brazilian Program for Biodiversity Research (PPBio) Information System. *Biodiversity & Ecology* 4: 265–274

Viana, P.L., **Carvalho, F.A.** & Reis, I. (2010) Tetrameristaceae (Magnoliophyta: Ericales): Primeiro Registro da Família Para o Brasil. *Revista Brasil. Bot.* 33(2): 375–378

Carvalho, F.A., Costa, F.R.C. & Salino, A. (2007) Determinantes da estrutura da comunidade de pteridófitas na BR 319, interflúvio Purus-Madeira Amazonas, Brasil. *Revista Brasileira de Biociências* 5(2): 1074–1076, Porto Alegre, RS

Book chapters

Carvalho FA (2015) Caricaceae. *In:* Davidse G, Sousa M, Knapp S, Chiang Cabrera F & Ulloa Ulloa C (eds) Flora Mesoamericana Vol. 2, Parte 3: Saururaceae a Zygophyllaceae, Missouri Botanical Garden Press, Monsanto

Carvalho F.A., Renner S.S. (2013) The phylogeny of Caricaceae. *In:* Ming R., Moore P.H. (eds) Genetics and Genomics of Papaya. Springer, New York, pp. 81–82

Salino, A., **Carvalho, F.A.** (2005) Dryopteridaceae. In: Cavalcanti T.B., Ramos, A.E. (eds) Flora do Distrito Federal, Brasil. Embrapa Recursos Genéticos e Biotecnologia, Brasília. v.4, pp.137–143

Proceedings

Carvalho, F.A., Renner, S.S. Climatic niche divergence in old sister lineage splits of Caricaceae, but not young species pairs. In: 2nd BioSyst.EU meeting. Global systematics. 18–22 Feb 2013, Vienna, Austria. BioSyst. EU Abstract book, p. 11. **Oral presentation**

Carvalho, F.A., Renner, S.S. The papaya (*Carica papaya*) tree belongs in an herbaceous Mesoamerican clade. In: 7th International Congress of Systematic and Evolutionary Biology, 21–27 Feb 2011. Biosystematics Berlin Abstracts book, p. 39. **Oral presentation**

Carvalho, F. A., Hopkins, M.G.J., Viana, P.L. & Assunção, P.A. Uma metodologia rápida para descrever floras locais. In: International Scientific Conference "Amazon in Perspective: Integrated Science for a Sustainable Future". 19 Nov. 2008, Manaus, AM. **Poster**. Abstract ID: 439. Avaliable at http://www.lbaconferencia.org/cgibin/lbaconf_2008/conf08_ab_search.pl

Carvalho, F.A., Viana, P.L. & Reis, I. Tetrameristaceae: first record of the family in Brazil. In: International Scientific Conference "Amazon in Perspective: Integrated Science for a Sustainable Future, 19 Nov. 2008, Manaus, AM, Brazil. **Poster**. Abstract ID: 499. Avaliable at http://www.lbaconferencia.org/cgibin/lbaconf_2008/conf08_ab_search.pl

Carvalho, F.A., Castilho, C.V., Costa, F.R. Rede de Parcelas Permanentes da Reserva Biológica do Uatumã: Avaliação da Estrutura da Vegetação. In: International Scientific Conference "Amazon in Perspective: Integrated Science for a Sustainable Future. 20 Nov. 2008, Manaus, AM, Brazil. **Poster**. Abstract ID: 616. Avaliable at http://www.lbaconferencia.org/cgibin/lbaconf_2008/conf08_ab_search.pl

Carvalho, F.A., Costa, F.R.C. & Salino, A. Diversidade beta no interflúvio Purus-Madeira: determinantes da estrutura das comunidades de Marantaceae, Araceae e Pteridófitas na BR 319, Amazonas, Brasil. In: II Mostra da Pós-Graduação do Amazonas. Manaus, AM, Brazil. CD of abstracts. October 2006. **Oral presentation**

Añez, R.B.S. & **Carvalho, F.A.** Morfo - Anatomia foliar e aspectos da Biologia Floras de Rhodostemonodaphne crenaticupula Martiñan (Lauraceae) na campinarana, Reserva da Campina, Manaus, AM. In: 56. Congresso Nacional de Botânica, Curitiba, PR, Brazil. CD of abstracts, July 2005. **Poster**

Salino, A., Almeida, T.E., Mota, N.F.O., **Carvalho, F.A.**, Mota, R.C., Lombardi, J.A., Stehman, J.R. (2005) Pteridófitas de fragmentos de Mata Atlântica no Vale do Jequitinhonha no nordeste de Minas Gerais, Brasil. In: 56. Congresso Nacional de Botânica. Curitiba, PR, Brazil. CD of abstracts, July 2005. **Poster**

Salino, A., Dittrich, V.O. & **Carvalho, F.A.** Pteridófitas do Parque Estadual de Intervales, Estado de São Paulo, Brasil. In: 55. Congresso Nacional de Botânica. Viçosa, MG, Brazil. CD of abstracts, September 2004. **Poster**

Carvalho, F.A., Salino, A., Viana, P.L. & Mota, R.C. Pteridófitas do Parque Municipal Natural ribeirão do Campo, Conceição do Mato Dentro, Estado de Minas Gerais, Brazil. In: 54. Congresso Nacional de Botânica. Belém, PA, Brazil. CD of abstracts, July 2003. **Poster**

Carvalho, F.A., Morais, P.O., Dittrich, V.O., Salino, A., Teixeira, L.C.R.S. Pteridófitas do Parque Estadual da Serra do Mar, São Paulo, Brasil. In: 54. Congresso Nacional de Botânica. Belém, PA, Brazil. CD of abstracts. July 2003. **Poster**

Carvalho, F.A., Teixeira, L.C.R.S, Salino. A. Levantamento das pteridófitas de cinco áreas da APA Sul, Minas Gerais, Brasil. (2002) In: 24. Encontro regional de Botânicos. Ilhéus, BA, Brazil. April 2002. **Poster**

Printed in the United States
By Bookmasters